HEINEMANN MODULAR MATHEMATICS
for
LONDON AS AND A-LEVEL
Mechanics 1

Jean Littlewood   John Hebborn   Fred Norton

Heinemann Educational Publishers,
a division of Heinemann Publishers (Oxford) Ltd,
Halley Court, Jordan Hill, Oxford, OX2 8EJ

OXFORD LONDON EDINBURGH MADRID ATHENS BOLOGNA
PARIS MELBOURNE SYDNEY AUCKLAND SINGAPORE
TOKYO IBADAN NAIROBI HARARE GABORONE
PORTSMOUTH NH (USA)

First published 1994

95 96 10 9 8 7 6 5 4 3

ISBN 0 435 51803 8

Original design by Geoffrey Wadsley; additional design work by Jim Turner

Typeset and illustrated by TecSet Limited, Wallington, Surrey.

Printed in Great Britain by The Bath Press, Avon

Acknowledgements:

The publisher's and authors' thanks are due to the University of London
Examinations and Assessment Council (ULEAC) for permission to reproduce
questions from past examination papers. These are marked with an [L].
   The answers have been provided by the authors and are not the responsibility
of the examining board.

About this book

This book is designed to provide you with the best preparation possible for your London Modular Mathematics M1 examination. The series authors are examiners and exam moderators themselves and have a good understanding of the exam board's requirements.

Finding your way around

To help to find your way around when you are studying and revising use the:

- **edge marks** (shown on the front page) – these help you to get to the right chapter quickly;
- **contents list** – this lists the headings that identify key syllabus ideas covered in the book so you can turn straight to them;
- **index** – if you need to find a topic the **bold** number shows where to find the main entry on a topic.

Remembering key ideas

We have provided clear explanations of the key ideas and techniques you need throughout the book. Key ideas you need to remember are listed in a **summary of key points** at the end of each chapter and marked like this in the chapters:

■ $$v = u + at$$

Exercises and exam questions

In this book questions are carefully graded so they increase in difficulty and gradually bring you up to exam standard.

- **past exam questions** are marked with an L;
- **review exercises** on pages 81 and 217 help you practise answering questions from several areas of mathematics at once, as in the real exam;
- **exam style practice paper** – this is designed to help you prepare for the exam itself;
- **answers** are included at the end of the book – use them to check your work.

Contents

3 Kinematics of a particle

4 Statics of a particle

5 Dynamics of a particle moving in a straight line

6 Moments

Basic mathematical techniques

1 Trigonometry

Sine rule

For this triangle:

$$\frac{a}{\sin A} = \frac{b}{\sin B} = \frac{c}{\sin C}$$

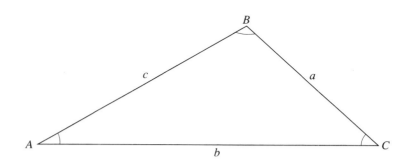

Example

In the triangle ABC, $\angle A = 65°$, $\angle B = 40°$ and $c = 5$ cm.

Find the size of the other angle and the lengths of the other sides.

As: $$\angle A + \angle B + \angle C = 180°$$

then: $$\angle C = 180° - 65° - 40°$$
$$= 75°$$

Using the sine rule gives: $$\frac{a}{\sin 65°} = \frac{b}{\sin 40°} = \frac{5}{\sin 75°}$$

So: $$a = \frac{5 \sin 65°}{\sin 75°} = 4.69$$

and: $$b = \frac{5 \sin 40°}{\sin 75°} = 3.33$$

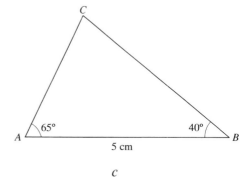

So: $\angle C$ is 75°, side a is 4.69 cm and side b is 3.33 cm.

Cosine rule

For the triangle as shown:

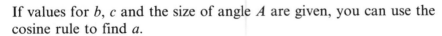

$$a^2 = b^2 + c^2 - 2bc \, \cos A$$

If values for a, b and c are given you can use the cosine rule to find the size of angle A.

If values for b, c and the size of angle A are given, you can use the cosine rule to find a.

For other combinations of given values, use the sine rule as it is easier to apply.

Example

Given $a = 4$ cm, $b = 4.5$ cm, $c = 5$ cm, find the size of angle A.

Rearranging the cosine rule gives:

$$\cos A = \frac{b^2 + c^2 - a^2}{2bc}$$

$$= \frac{(4.5)^2 + (5)^2 - (4)^2}{(2) \times (4.5) \times (5)}$$

So:
$$\angle A = 49.5°$$

Example

Given $\angle A = 121°$, $b = 6.9$ cm, $c = 4.7$ cm find a.

$$a^2 = b^2 + c^2 - 2bc \, \cos A$$

$$= (6.9)^2 + (4.7)^2 - 2 \times (6.9) \times (4.7) \times \cos 121°$$

$$a = 10.2 \, \text{cm}$$

2 Calculus

Differentiation

If $y = x^p$, where p is a rational number, then $\dfrac{\mathrm{d}y}{\mathrm{d}x} = px^{p-1}$

Example

If:
$$y = 5x^2 + 6x + 5 + 9x^{-2}, \quad x > 0$$

then:
$$\frac{\mathrm{d}y}{\mathrm{d}x} = 10x + 6 + 0 - 18x^{-3}, \quad x > 0$$

Integration

If:
$$y = x^p \quad (p \neq -1)$$

then:
$$\int y \, dx = \frac{x^{p+1}}{p+1} + c$$

where c is an arbitrary constant.

Example

If:
$$y = 5x^2 + 6x + 5 + 9x^{-2}, \quad x > 0$$

then:
$$\int y \, dx = \frac{5x^3}{3} + \frac{6x^2}{2} + 5x + \frac{9x^{-1}}{(-1)} + c$$
$$= \frac{5}{3}x^3 + 3x^2 + 5x - 9x^{-1} + c$$

Quadratic equations

The solutions of the quadratic equation $ax^2 + bx + c = 0$ are given by:

$$x = \frac{-b \pm \sqrt{(b^2 - 4ac)}}{2a}$$

These roots are often found by factorising the quadratic expression $ax^2 + bx + c$.

Example

Solve the equation $2x^2 + 3x - 2 = 0$.

Solutions are:
$$x = \frac{-3 \pm \sqrt{(9 + 16)}}{4}$$
$$= \frac{-3 \pm 5}{4}$$
$$= -2 \quad \text{or} \quad \tfrac{1}{2}$$

Alternatively: $(x + 2)(2x - 1) = 0$

So: $x + 2 = 0,$

or: $2x - 1 = 0$

Hence: $x = -2$

or: $x = \tfrac{1}{2}.$

3 Accuracy

Unless otherwise stated all numerical answers should be given to 3 significant figures. Before you arrive at your final answer, use your calculator to calculate using the maximum number of figures and then approximate the final answer to 3 significant figures. Do not round-up any of the intermediate answers in a complex calculation because this will give an inaccurate final answer.

Mathematical models in mechanics

1

1.1 What is mathematical modelling?

Physicists and engineers have used mathematics to solve problems in the real world for more than 200 years. Nowadays it is also used to solve problems in many other fields such as biology, economics, geography, medicine and psychology. Examples of problems which may be solved using mathematics include:

- estimating the height of the leaning tower of Pisa–without climbing it
- estimating the width of a river–without crossing it
- predicting the population of China in the year 2000–without waiting until then
- predicting the effect of a 30 per cent reduction in income tax–without actually reducing the rate
- estimating the volume of blood inside someone's body–without bleeding him to death!.

These problems and many others may be solved using a process called **mathematical modelling**. Mathematical modelling involves:

- **translating** a real world problem into a mathematical problem or **model**;
- **solving** the mathematical problem;
- **interpreting** the solution in terms of the real world.

The process can be shown as a diagram:

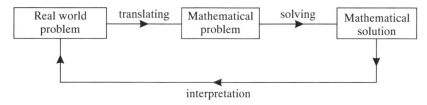

Real world problems can't often be translated perfectly into mathematical problems. Even when they can, the resulting mathematical problem may be so complex that it is not possible to solve it. So it is often necessary to simplify a real world problem into one which produces a solvable mathematical problem by concentrating on its essential features and ignoring the rest.

Sometimes the amount of simplification required may seem drastic. For example, to model the motion of the planets, the planets and the Sun can be considered to be point masses and their size and structure ignored. Simplifications like this may be justified if the resulting mathematical model produces predictions for the motions of the planets that match up to the motions that are actually observed. If the model does not produce reliable predictions it may need to be changed in some way, usually by considering the assumptions made in the simplification. The modelling process can be shown in a diagram like this:

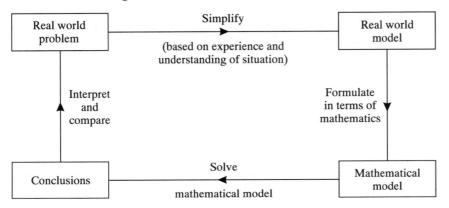

Example 1

A student is asked to estimate her travelling expenses when invited to an interview. She models the situation in a very simple way. First she looks up the distance d km in her road atlas. She finds from her car manual that her car should do k km per litre. She knows from petrol station prices that the cost of unleaded petrol is £c per litre.

Based on this data she estimates her travelling expenses to be

$$£\left(\frac{d}{k}\right)c.$$

This is a simple model which ignores many factors that may affect the number of km per litre her car will actually travel: the traffic conditions, possible roadworks, the types of roads she travels on, the time of day she travels – rush hour or off peak. It also ignores other running costs of her car.

Using her experience she may then improve or refine her model, for example by changing the value of k or building in some of the factors neglected in her simple model.

Example 2

Here is a much more complex problem – how can we estimate the rate at which a country's population will grow? Changes in the way resources are used need to be planned well ahead, so governments need good estimates of how the population is likely to change. The real world problem is to explain how populations change, and to create a model to predict future changes.

The first model was put forward by Thomas Malthus as long ago as 1798. He suggested that the rate of change of a population of size N people is kN, where k is a positive constant. The mathematical solution of this problem produces the graph shown here.

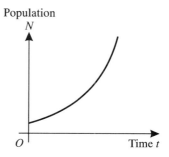

Population N

Time t

Malthus's model predicts an unlimited growth in population, which is unlikely to occur. In 1837 Pierre Verhulst refined this model by looking at factors which had been ignored in it – in particular factors that might limit population growth. He proposed that the rate of change of a population N is:

$$kN\left(1 - \frac{N}{M}\right)$$

This is based on the assumption that there is an upper limit M to the size of population which can be sustained. The mathematical solution of this problem produces the graph shown here.

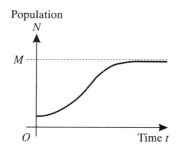

Population N

M

Time t

This prediction fitted the population data for the USA from 1810 to 1930. However, it does not agree with population data since then, so further refinements to the model are required.

Exercise 1A

Briefly discuss how you would set up a mathematical model to solve each of the following problems. Include details of which quantities are relevant, any assumptions you have made and any refinements you could make to improve your model.

1 A family of four with two adults and two children aged 8 and 10 wishes to travel to London for a sporting event and to keep their costs to a minimum. Should they travel by car, by coach, by train or some combination of these?

2 A family needs to estimate its household bills for the next three months to decide if it can afford a holiday. Do this for a family like yours.

3 A lecturer has a choice of three routes to drive from college to home. Which one should he choose in rush hour to keep his journey time as short as possible?

4 A woman uses her car for business. She needs to estimate the annual cost of running her car. How can she do this?

5 A company is organising a one-day conference which involves certain expenses. Last year 100 people attended. What conference fee should the company charge each delegate so that it covers its expenses?

6 A new supermarket has opened in a town 10 miles away from where the Browns live. The prices appear to be cheaper than those of their local corner shop. Help the Browns decide whether it will be cheaper to shop locally or at the new supermarket.

7 A family of four adults decides to go to the Channel Islands for a holiday. They agree that they need the use of a car while they are there. They wish to keep the cost to a minimum. The possibilities are (a) fly and hire a car (b) go by boat and hire a car or take their own (c) go on the new catamaran and hire a car or take their own.
Which option should they choose?

1.2 Mathematical models in mechanics

Mathematical modelling involves simplifying a real world problem to produce a model which can be solved mathematically. In mechanics a range of mathematical models has been developed over the years to describe the effects of forces on various types of objects in practical situations.

The basic work on mechanics in this book deals with simple objects and only the most important forces acting on them, allowing you to produce mathematical models which you can solve.

Basic terminology for mechanics

This section introduces the essential terminology you need to know for modelling in mechanics.

- A **particle** is a body whose dimensions are so small compared with the other lengths involved that its position in space can be represented by a single point. For example, in considering the motion of the Earth relative to the Sun you may represent the Earth and the Sun by particles.

- A **lamina** is a flat object whose thickness is small compared with its width and length. For example, a piece of card, a sheet of paper or a thin metal sheet may be represented by laminae (plural of lamina).

- A **uniform lamina** is one in which equal areas of the lamina have equal masses. This is clearly the case when the whole of the lamina is made of the same material.

- A **rigid body** is an object made up of particles, all of which remain at the same fixed distances from one another whether the object is at rest or in motion. For example, a rigid beam is assumed to keep its shape when acted on by forces. When a billiard ball strikes the edge of a snooker table you may assume that it does not change its shape, so it can be represented by a rigid body.

- A **rod** is an object all of whose mass is concentrated along a line. It is assumed to have length only, and its width and breadth are neglected. A broom pole may, for example, be modelled by a rod.

- A **uniform rod** is one in which equal lengths have equal masses.

- A **light object** is one whose mass is so small compared with other masses being considered that the mass may be considered to be zero. A **light string** is one example. If an object is suspended by a light string, the mass of the string may be ignored. A **light rod** is another example. If two particles are joined by a light rod these particles remain the same distance apart. The mass of the rod may be neglected.

- An **inextensible string** or **inelastic string** is a string whose length remains the same whether motion is taking place or not. While all real strings are elastic to some extent, in many problems the extension is so small compared to the other lengths under consideration that it may be ignored.

- A **smooth surface** is one which offers so little frictional resistance to the motion of a body sliding across it that the friction may be ignored. A sheet of ice is an example of a smooth surface commonly found in the real world.

- A **smooth pulley** is one with no friction in its bearings.

- A **plane surface** is a completely flat surface. Walls, floors and tables are usually modelled by plane surfaces.

- The **Earth's surface** is usually modelled by a horizontal plane surface. (You may also model slopes by inclined planes.)

An example of modelling in mechanics

As an example of modelling in mechanics, think about the motion of a cricket ball thrown through the air. In the simplest possible model the following assumptions are made:

The ball may be modelled by a particle	(1)
The acceleration due to gravity is constant	(2)
The motion of the ball takes place in a vertical plane	(3)
The only force acting on the ball is its weight.	(4)

Assumption (1) ignores the size and shape of the ball – sometimes called the **particle model**.

Assumption (2) ignores the variation of gravity with position on the Earth's surface and height above sea level.

Assumption (3) ignores any sideways movement due to crosswinds or spin.

Assumption (4) ignores the effect of any air resistance.

A detailed solution for this model is given on page 6. At this point just consider how well the model matches the real world. In the model the path of the ball is a symmetrical curve called a parabola shown here:

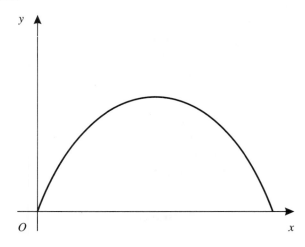

However in the real world the ball's path looks like this:

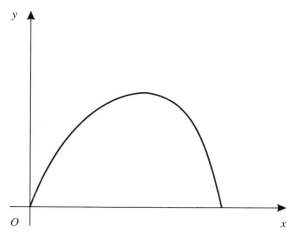

To get a better match to the real world the model needs to be refined. If account is taken of air resistance, which depends on the speed of the ball, the new model is a better match to the real world situation.

Exercise 1B

1 Suggest a simple mathematical model for each of the following problems.

(a) A brick of mass 1 kg is attached to one end of a rope. A boy picks up the other end of the rope and swings the brick so that its path of motion is a vertical circle. Find how the tension in the rope is related to the speed of the brick.

(b) A book of mass 0.5 kg is released from rest on the lid of a polished wood desk inclined at 20° to the horizontal. Find the distance travelled by the book in 5 seconds.

(c) In a game of snooker the white ball strikes a red ball which is at rest. Find how the speed of the red ball, after the collision, is related to the speed of the white ball which strikes it.

(d) A see-saw is made up of a plank of length 4 m balanced on a support at its centre. Andrew has a mass of 25 kg and his sister Brenda has a mass of 20 kg. Andrew sits at one end of the plank. Where should Brenda sit if the see-saw is to balance?

(e) A man needs to climb to the top of a ladder to carry out some repairs. The ladder rests on a rough plane against a rough wall. If you know the man's weight how could you find the maximum inclination of the ladder for it still to be safe to climb?

2 In each of the following a physical situation is given together with a mathematical model. Decide if the model is a good one and suggest ways it could be refined.

(a) A cricket ball is thrown vertically upwards. Given its initial speed find how long it takes to return to its initial position.

Model: A particle moves vertically under constant gravity with no air resistance.

(b) In a game of ice hockey a puck is hit across an ice rink. Given the initial speed find how far it travels in 6 seconds.

Model: A particle moves on a smooth horizontal surface.

(c) A piece of equipment used in a gym consists of two metal spheres each of mass 20 kg and radius 0.1 m joined by a metal rod of mass 0.5 kg and length 2 m. It is dropped from a height of 2 m with the rod horizontal. Determine the speed with which the equipment reaches the ground.

Model: Two particles of mass 20 kg are joined by a light rod of length 2.2 m. The system starts from rest with the rod horizontal and falls freely under gravity.

(d) A builder attaches a large bucket to one end of a rope which passes over a pulley and he uses this to hoist materials to a second floor window. Find the maximum weight the bucket will carry.

Model: A light inextensible string passes over a smooth pulley with a particle attached at the end of the string. Find the tension in the string for a given mass.

(e) In pole-vaulting an athlete uses a glass fibre pole to attempt to clear a horizontal bar. How does his initial speed affect the height he can clear?

Model: A particle of mass equal to that of the athlete is attached at the end of a light rigid rod. The particle moves freely under gravity on leaving the rod.

Vectors and their application in mechanics

2

2.1 Vector and scalar quantities

When applying mathematics to mechanical systems you will often need to carry out calculations involving quantities such as:

- the mass of a particle (for example 10 g)
- the distances between two points on a table (for example 0.5 m)
- the time taken by a runner to complete a 100 m race (say 10 s).

Quantities like these are called **scalars**. Each one is completely described by a number and the appropriate unit of measurement.

It is not too difficult, however, to think of quantities which are *not* completely characterised by a single number.

Suppose you are told that a particle has moved a distance of 5 m from a point O. You do not have enough information to find the final position of the particle. As well as the distance it has moved you need to know the *direction* in which the particle has moved. If the final position of the particle is the point A, then you can say that the **displacement** of the particle is OA. To specify this displacement you need the **length** of OA and the **direction** of $\overrightarrow{OA}$. The arrow shows that the particle has moved from O to A.

Suppose you now are told that a force of a certain size is applied to a particle by means of a string attached to the particle. Again, this information is not sufficient for you to be able to say what happens to the particle. In addition to knowing the **magnitude** or size of the force, you need to know the **direction** in which this force acts.

In both of these examples, to specify the quantities completely you needed to give a number and also a direction.

- **A vector is defined to be any quantity which is completely specified by a scalar magnitude and a direction.**

Displacements and forces are therefore **vectors**. Later in this chapter you will see that velocities and accelerations are also vectors.

Example 1

A man starts at the point O and walks 4 km east to the point A and then 3 km north to the point B. Find his final displacement from O.

His journey can be shown by drawing a diagram. This is an important first step in a problem of this type.
Using Pythagoras' Theorem:

$$OB^2 = OA^2 + AB^2$$
$$= 4^2 + 3^2$$
$$= 16 + 9 = 25$$
$$OB = 5$$

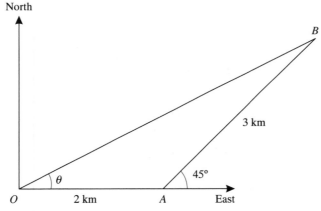

The length OB is 5 km.

The *displacement* of the man is given by OB which is of length 5 km.

The *distance* he walked is given by:

$$OA + AB = 7 \text{ km}$$

The displacement is represented by the directed straight line $\overrightarrow{OB}$.

Example 2

A man walks 2 km due east from O to A and then 3 km in a north-east direction from A to B. Find the distance of B from O and describe the displacement OB.

First draw a diagram representing the man's journey.

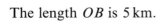

Using the cosine rule gives:

$$OB^2 = OA^2 + AB^2 - 2 \times OA \times AB \cos B\hat{A}O$$

As: $\qquad \angle BAO = (180 - 45)^\circ = 135^\circ$

$$OB^2 = 2^2 + 3^2 - 2 \times 2 \times 3 \cos 135^\circ$$

So: $\qquad OB = 4.64 \text{ km}$

To find the direction of the displacement OB you need to find $\angle BOA$.

Let $\angle BOA = \theta$. Using the sine rule gives:

$$\frac{OB}{\sin 135°} = \frac{AB}{\sin \theta}$$

$$\frac{4.64}{\sin 135°} = \frac{3}{\sin \theta}$$

$$\sin \theta = \frac{3 \sin 135°}{4.64}$$

So: $\qquad\qquad\qquad \theta = 27.2°$

Usually when you specify directions you refer them to the north–south line. The angle OB makes with the north line is $(90 - 27.2)° = 62.8°$. So the displacement $\overrightarrow{OB}$ is in the direction N 62.8° E.

Exercise 2A

1 A boy walks x km due east then y km due north. Calculate the total distance he has walked and his displacement from his starting point when:
(a) $x = 5$, $y = 12$ (b) $x = 12$, $y = 5$ (c) $x = 10$, $y = 7$
(d) $x = 1$, $y = 16$ (e) $x = 10$, $y = -27$.
What is the significance of the negative sign in (e)?

2 A girl walks x km due east then z km north-east. Calculate the total distance she has walked and her displacement from her starting point when:
(a) $x = 2$, $z = 5$ (b) $x = 3$, $z = 4$ (c) $x = 5$, $z = 2$
(d) $x = 1$, $z = 6$ (e) $x = 8$, $z = -1$.

3 In a regatta a yacht sails x km due east from a marker O then y km due north to a buoy B. Calculate the magnitude and direction of the displacement vector $\overrightarrow{OB}$ when:
(a) $x = 5$, $y = 8$ (b) $x = 8$, $y = 5$ (c) $x = 10$, $y = 16$.

4 In a desert exercise a tank travels x km on a bearing of $050°$ from a base O then y km on a bearing of $140°$ to a bunker B. Calculate the magnitude and direction of the displacement vector $\overrightarrow{OB}$ when:
(a) $x = 7$, $y = 6$ (b) $x = 6$, $y = 7$ (c) $x = 12$, $y = 14$.

5 When orienteering, Jack walks x km on a bearing of 100° from his starting point O then y km on a bearing of 200° to a hut H. Calculate the magnitude and direction of the displacement vector $\overrightarrow{OH}$ when:

(a) $x = 6$, $y = 7$ (b) $x = 6$, $y = 6$ (c) $x = 6$, $y = 10$.

6 Jill is swimming across a river which has a constant current of $k\,\mathrm{m\,s^{-1}}$. If Jill heads directly across the river with speed $v\,\mathrm{m\,s^{-1}}$, find her resultant speed and the angle her path makes with a line directly across the river when:

(a) $k = 0.5$, $v = 1$ (b) $k = 0.5$, $v = 1.5$ (c) $k = 2$, $v = 1$.

2.2 Properties of vectors

Representing vectors

Any vector can be represented by a **directed line segment**. The direction of the line segment is that of the vector, and its length represents the magnitude of the vector.

So $\overrightarrow{PQ}$ represents the vector with magnitude and direction given by the line segment joining P and Q.

The magnitude of the vector is written $|\overrightarrow{PQ}|$.

Equal vectors

Two vectors are **equal** if and only if they have both the **same magnitude** and the **same direction**.

In the diagram, PQ and LM are parallel and also equal in length so $\overrightarrow{PQ} = \overrightarrow{LM}$.

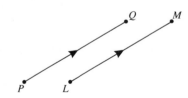

Null vector or zero vector

The magnitude of a vector is usually positive. It is never negative, but it may be zero. In this case it is called the **zero vector** or **null vector**.

Adding vectors

Suppose a runner goes from O to A and then from A to B and a second runner goes directly from O to B.

As the result of these two journeys is the same you can write:

$$\overrightarrow{OA} + \overrightarrow{AB} = \overrightarrow{OB}$$

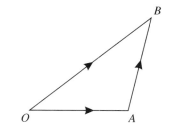

The displacement $\overrightarrow{OB}$ is the **sum** or **resultant** of the displacements $\overrightarrow{OA}$ and $\overrightarrow{AB}$. This rule is called the **triangle law of addition**.

■ **All vectors represented by directed line segments can be manipulated in the same way as displacement vectors — and can be added using the triangle law of addition.**

Another way of adding vectors is to use the **parallelogram rule**. If you complete the parallelogram, with sides of $\overrightarrow{OA}$ and $\overrightarrow{AB}$, you get the following:

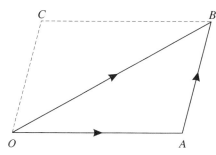

The resultant of $\overrightarrow{OA}$ and $\overrightarrow{AB}$ is then represented by the **diagonal** $\overrightarrow{OB}$ of the parallelogram OABC.

Vectors that are related to each other

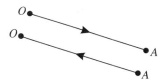

Think of a journey from O to A followed by the return journey from A to O. The result of these two journeys is a zero displacement (you end up back where you started). So you can write:

$$\overrightarrow{OA} = -\overrightarrow{AO}$$

You can see that the **order of the letters and the direction of the arrow are important**. $\overrightarrow{OA}$ is not the same as $\overrightarrow{AO}$.

The two displacements $\overrightarrow{PQ}$ and $\overrightarrow{LM}$ are parallel but of different magnitudes.

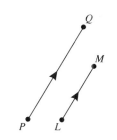

You can write this as:

$$\overrightarrow{PQ} = k\overrightarrow{LM}$$

where k is a scalar that represents the ratio of the magnitudes $|\overrightarrow{PQ}|$ and $|\overrightarrow{LM}|$.

Another type of vector notation

It is often convenient to use an alternative notation in which a vector $\overrightarrow{OA}$ is represented by a single letter, usually written alongside the vector in a diagram. When this is done the letter is always printed in **bold type**. In manuscript it should be written with a wavy line underneath for example $\underset{\sim}{a}$. The magnitude of the vector is usually represented by the same letter but printed in *italic*. So:

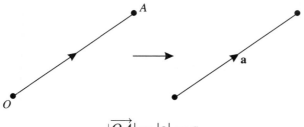

And:
$$|\overrightarrow{OA}| = |\mathbf{a}| = a$$

Using this notation the triangle law becomes:

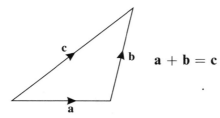

$$\mathbf{a} + \mathbf{b} = \mathbf{c}$$

Also, in this notation $-\mathbf{b}$ has the same magnitude as $\mathbf{b}$ but is in the opposite direction:

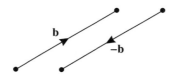

Example 3

The diagram shows a parallelogram $OACB$ with $\overrightarrow{OA} = \mathbf{a}$ and $\overrightarrow{OB} = \mathbf{b}$. The point D is the mid-point of AC. Express the following vectors in terms of $\mathbf{a}$ and $\mathbf{b}$:

(a) $\overrightarrow{BC}$ (b) $\overrightarrow{AC}$ (c) $\overrightarrow{CA}$ (d) $\overrightarrow{OC}$ (e) $\overrightarrow{AB}$ (f) $\overrightarrow{OD}$.

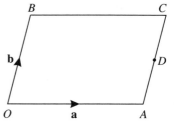

(a) $\overrightarrow{BC}$ is in the same direction as $\overrightarrow{OA}$ and of the same length, because $OACB$ is a parallelogram.

So:
$$\overrightarrow{BC} = \overrightarrow{OA} = \mathbf{a}$$

(b) Similarly, $\overrightarrow{AC}$ is in the same direction as $\overrightarrow{OB}$ and of the same length.

So:
$$\overrightarrow{AC} = \overrightarrow{OB} = \mathbf{b}$$

(c) The vector $\overrightarrow{CA}$ is in the opposite direction to $\overrightarrow{AC}$ but of the same length.

So:
$$\overrightarrow{CA} = -\overrightarrow{AC} = -\mathbf{b}$$

(d) To find $\overrightarrow{OC}$, consider the triangle OAC.
By the triangle law:
$$\overrightarrow{OC} = \overrightarrow{OA} + \overrightarrow{AC}$$
$$= \overrightarrow{OA} + \overrightarrow{OB}$$
$$= \mathbf{a} + \mathbf{b}$$

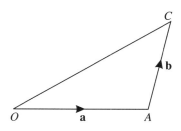

(e) To find $\overrightarrow{AB}$, consider the triangle OAB.
By the triangle law:
$$\overrightarrow{OA} + \overrightarrow{AB} = \overrightarrow{OB}$$

Or:
$$\overrightarrow{AB} = \overrightarrow{OB} - \overrightarrow{OA}$$
$$= \mathbf{b} - \mathbf{a}$$

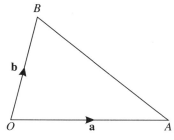

(f) To find $\overrightarrow{OD}$, consider the triangle OAD.
You must first find $\overrightarrow{AD}$. $\overrightarrow{AD}$ is parallel to $\overrightarrow{AC}$ and half its length as D is the mid-point of $\overrightarrow{AC}$.

So:
$$\overrightarrow{AD} = \tfrac{1}{2}\mathbf{b}$$

By the triangle law:
$$\overrightarrow{OD} = \overrightarrow{OA} + \overrightarrow{AD}$$
$$= \mathbf{a} + \tfrac{1}{2}\mathbf{b}$$

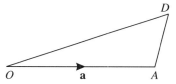

Exercise 2B

1

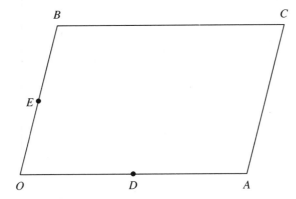

In the parallelogram $OACB$, D is the mid-point of OA and E is the mid-point of OB. Vectors **a** and **b** are represented in magnitude and direction by OA and OB. Find in terms of **a** and **b** the vectors represented in magnitude and direction by:

(a) OD (b) EO (c) AB (d) AC (e) CA

(f) BE (g) DE (h) OC (i) CO (j) DC.

2

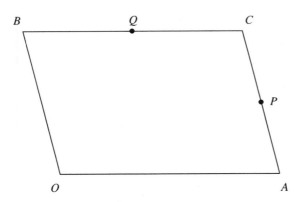

In the parallelogram $OACB$, P is the mid-point of AC and Q is the mid-point of BC. Vectors **a** and **b** are represented in magnitude and direction by OA and OB. Find in terms of **a** and **b** the vectors represented in magnitude and direction by:

(a) AC (b) CB (c) OP (d) AQ (e) QA

(f) QB (g) QP (h) PQ (i) AB (j) PB.

3

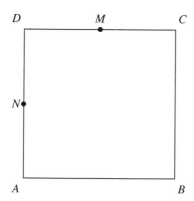

In the square $ABCD$, M is the mid-point of CD and N the mid-point of AD. Vectors **a** and **b** are represented in magnitude and direction by AB and AD. Find in terms of **a** and **b** the vectors represented in magnitude and direction by:

(a) AC (b) BD (c) BM (d) MA (e) AM
(f) NM (g) CN (h) MN (i) NA (j) CB.

4

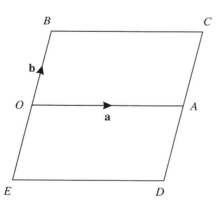

$OACB$ and $OADE$ are two congruent parallelograms. OA and OB represent in magnitude, direction and line of action the vectors **a** and **b**. Find the directed line segments which represent in magnitude, direction and line of action:

    (a) $-\mathbf{a}$ (b) $-\mathbf{b}$ (c) $\mathbf{a} + \mathbf{b}$ (d) $\mathbf{a} - \mathbf{b}$ (e) $\mathbf{b} - \mathbf{a}$.

5

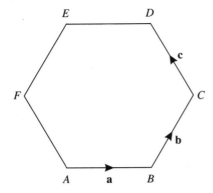

The sides *AB*, *BC* and *CD* of the regular hexagon *ABCDEF* represent in magnitude and direction the vectors **a**, **b**, and **c** respectively. Find in terms of **a**, **b** and **c** the vectors represented in magnitude and direction by:

(a) *DE*   (b) *FE*   (c) *AF*   (d) *AC*   (e) *AD*
(f) *AE*   (g) *BF*   (h) *EC*   (i) *DF*   (j) *DB*.

2.3 Cartesian unit vectors and components

A **unit vector** is a vector with magnitude of 1 unit.
Chapter 4 of Book P1 deals with coordinate geometry. Particularly useful unit vectors are those that lie in the directions of the positive *x*-axis and the positive *y*-axis (the cartesian axes) discussed in that chapter.

i, j notation

The unit vectors along the cartesian axes are usually denoted by **i** and **j** respectively:

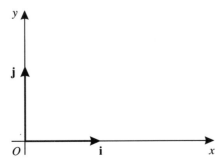

Example 4

A vector $(3\mathbf{i} + 4\mathbf{j})$ units represents a displacement of:

3 units in the direction of the unit vector $\mathbf{i}$,
4 units in the direction of the unit vector $\mathbf{j}$.

By the triangle law of addition this is just the vector $\mathbf{R}$ which completes the triangle and so:

$$\mathbf{R} = 3\mathbf{i} + 4\mathbf{j}$$

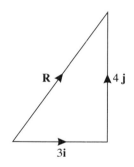

The vector $\mathbf{R}$ is written in the $\mathbf{i}, \mathbf{j}$ notation. The advantages of the $\mathbf{i}, \mathbf{j}$ notation will now be shown.

Adding vectors in i, j notation

When vectors are given in terms of unit vectors $\mathbf{i}$ and $\mathbf{j}$, you can add them together by adding the terms involving $\mathbf{i}$ and the terms involving $\mathbf{j}$ separately.

Example 5

Given $\mathbf{a} = 2\mathbf{i} + 3\mathbf{j}$ and $\mathbf{b} = (4\mathbf{i} - 2\mathbf{j})$ find $\mathbf{a} + \mathbf{b}$ in terms of $\mathbf{i}$ and $\mathbf{j}$.

Adding $\mathbf{a}$ and $\mathbf{b}$ gives:
$$\begin{aligned}
\mathbf{a} + \mathbf{b} &= (2\mathbf{i} + 3\mathbf{j}) + (4\mathbf{i} - 2\mathbf{j}) \\
&= (2\mathbf{i} + 4\mathbf{i}) + (3\mathbf{j} - 2\mathbf{j}) \\
&= 6\mathbf{i} + \mathbf{j}
\end{aligned}$$

This is shown in the diagram:

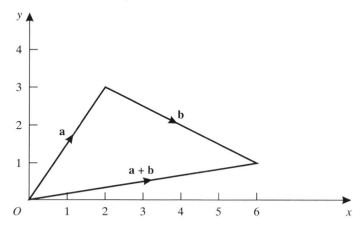

Subtracting vectors in i, j notation

You can subtract vectors given in the $\mathbf{i}, \mathbf{j}$ notation in a similar way, by subtracting the $\mathbf{i}$ parts and $\mathbf{j}$ parts separately.

Example 6

Given $\mathbf{a} = 6\mathbf{i} + 4\mathbf{j}$ and $\mathbf{b} = 4\mathbf{i} - 3\mathbf{j}$, find $\mathbf{a} - \mathbf{b}$ in terms of $\mathbf{i}$ and $\mathbf{j}$.

Subtracting $\mathbf{b}$ from $\mathbf{a}$ gives:
$$\begin{aligned} \mathbf{a} - \mathbf{b} &= (6\mathbf{i} + 4\mathbf{j}) - (4\mathbf{i} - 3\mathbf{j}) \\ &= (6\mathbf{i} - 4\mathbf{i}) + (4\mathbf{j} - (-)3\mathbf{j}) \\ &= 2\mathbf{i} + 7\mathbf{j} \end{aligned}$$

Magnitude of a vector in i, j notation

When a vector $\mathbf{R}$ is given in terms of the unit vectors $\mathbf{i}$ and $\mathbf{j}$ you can find its magnitude by using Pythagoras' Theorem. You can show that $\mathbf{R} = x\mathbf{i} + y\mathbf{j}$ as:

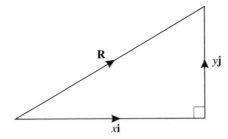

By Pythagoras' Theorem, $R^2 = x^2 + y^2$ and so the magnitude of $x\mathbf{i} + y\mathbf{j}$ is $\sqrt{(x^2 + y^2)}$.

Example 7

Find the unit vector in the direction of the vector $\mathbf{a} = 2\mathbf{i} + 3\mathbf{j}$.

The magnitude of $\mathbf{a}$ is:
$$|\mathbf{a}| = \sqrt{(2^2 + 3^2)} = \sqrt{13}$$

So the *unit* vector in the same direction as $\mathbf{a}$ is:
$$\frac{\mathbf{a}}{|\mathbf{a}|} = \frac{1}{\sqrt{13}}(2\mathbf{i} + 3\mathbf{j})$$

Equality of vectors

Suppose the two vectors $\mathbf{a}$ and $\mathbf{b}$ are given in the $\mathbf{i}$, $\mathbf{j}$ notation so that

$$\mathbf{a} = a_1\mathbf{i} + a_2\mathbf{j}$$
$$\mathbf{b} = b_1\mathbf{i} + b_2\mathbf{j}$$

These two vectors are equal only if:

$$a_1 = b_1 \text{ and } a_2 = b_2$$

On the other hand, if $a_1 = b_1$ and $a_2 = b_2$, then:

$$\mathbf{a} = \mathbf{b}$$

Example 8

Two vectors $\mathbf{a}$ and $\mathbf{b}$ are given by: $\mathbf{a} = (2 - x)\mathbf{i} + 4\mathbf{j}$
$$\mathbf{b} = \mathbf{i} + (6 - y)\mathbf{j}$$

where x and y are constants. Find x and y such that $\mathbf{a} = \mathbf{b}$.

If $\mathbf{a} = \mathbf{b}$ then: $(2 - x)\mathbf{i} + 4\mathbf{j} = \mathbf{i} + (6 - y)\mathbf{j}$

Equating the $\mathbf{i}$ parts gives: $2 - x = 1$

So: $x = 1$

Equating the $\mathbf{j}$ parts gives: $4 = 6 - y$

So: $y = 2$

Components of a vector

Any two vectors can be combined, using the triangle law of addition, into a single resultant (see p.17). It is possible and often useful to reverse this process; that is, to replace a vector by an equivalent pair of vectors. This process is called **resolving a vector into components**. This is very important in statics and will be considered further in chapter 4 with reference to forces. At this point we will consider only component vectors in the direction of the coordinate axes.

Look at the situation shown in the diagram.

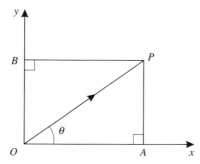

By the triangle law of addition:

$$\overrightarrow{OP} = \overrightarrow{OA} + \overrightarrow{AP}$$

But: $\overrightarrow{AP} = \overrightarrow{OB}$

So: $\overrightarrow{OP} = \overrightarrow{OA} + \overrightarrow{OB}$

By trigonometry:
$$\frac{OA}{OP} = \cos\theta \Rightarrow OA = OP\cos\theta$$
$$\frac{OB}{OP} = \sin\theta \Rightarrow OB = OP\sin\theta$$

So:
$$\overrightarrow{OA} = OP\cos\theta\,\mathbf{i}$$
$$\overrightarrow{OB} = OP\sin\theta\,\mathbf{j}$$

giving:
$$\overrightarrow{OP} = (OP\cos\theta)\,\mathbf{i} + (OP\sin\theta)\,\mathbf{j}$$

$OP\cos\theta$ and $OP\sin\theta$ are called the **components**, or **resolutes,** of $\overrightarrow{OP}$ along the coordinate axes. (They are sometimes called the **cartesian components.**)

Example 9

A displacement $\mathbf{R}$ is of magnitude of 16 km on a bearing of 030°. Take $\mathbf{i}$ as a unit vector due east and $\mathbf{j}$ as a unit vector due north and write $\mathbf{R}$ in the form $a\mathbf{i} + b\mathbf{j}$.

This information can be represented by a diagram:

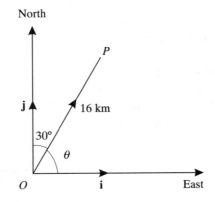

The angle θ is $90° - 30° = 60°$.

So:
$$\mathbf{R} = (R\cos\theta)\,\mathbf{i} + (\mathbf{R}\sin\theta)\,\mathbf{j}$$
$$= 16\cos 60°\,\mathbf{i} + 16\sin 60°\,\mathbf{j}$$

and:
$$\mathbf{R} = 8(\mathbf{i} + \sqrt{3}\mathbf{j})\text{ km}$$

using $\cos 60° = \frac{1}{2}$ and $\sin 60° = \frac{\sqrt{3}}{2}$, (see chapter 7, p. 142, of Book P1).

Example 10

The vector $\mathbf{V}$ is in the direction of the vector $8\mathbf{i} - 6\mathbf{j}$ and has magnitude 5. Find $\mathbf{V}$ in the form $a\mathbf{i} + b\mathbf{j}$.

$\mathbf{V}$ is in the direction of $(8\mathbf{i} - 6\mathbf{j})$, so:
$$\mathbf{V} = k(8\mathbf{i} - 6\mathbf{j})$$

where k is a positive constant that you must find.

As:
$$\mathbf{V} = 8k\mathbf{i} - 6k\mathbf{j}$$
$$|\mathbf{V}| = \sqrt{(64k^2 + 36k^2)} = 10k$$

But $\mathbf{V}$ has magnitude 5 and so:
$$10k = 5$$
$$k = \tfrac{1}{2}$$

Hence:
$$\mathbf{V} = \tfrac{1}{2}(8\mathbf{i} - 6\mathbf{i})$$
$$= 4\mathbf{i} - 3\mathbf{j}$$

In a case like this it is worth checking that the vector you have obtained *does* have the required magnitude.

Exercise 2C

In these questions, $\mathbf{i}$ and $\mathbf{j}$ are unit vectors in the direction of the positive x- and y-axes respectively.

1 Given that $\mathbf{a} = \mathbf{i} + 2\mathbf{j}$ and $\mathbf{b} = 2\mathbf{i} - \mathbf{j}$ find in terms of $\mathbf{i}$ and $\mathbf{j}$:
   (a) $\mathbf{a} + \mathbf{b}$ (b) $\mathbf{a} + 2\mathbf{b}$ (c) $2\mathbf{a} + \mathbf{b}$ (d) $\mathbf{a} - 2\mathbf{b}$
   (e) $2\mathbf{a} - \mathbf{b}$.

2 Given that $\mathbf{a} = 2\mathbf{i} + \mathbf{j}$ and $\mathbf{b} = \mathbf{i} + 3\mathbf{j}$ find (a) λ, if $\mathbf{a} + \lambda\mathbf{b}$ is parallel to the vector $\mathbf{i}$ (b) μ, if $\mu\mathbf{a} + \mathbf{b}$ is parallel to the vector $\mathbf{j}$.

3 Given that $\mathbf{a} = \mathbf{i} - 2\mathbf{j}$ and $\mathbf{b} = -3\mathbf{i} + \mathbf{j}$ find (a) λ, if $\mathbf{a} + \lambda\mathbf{b}$ is parallel to $-\mathbf{i} - 3\mathbf{j}$ (b) μ, if $\mu\mathbf{a} + \mathbf{b}$ is parallel to $2\mathbf{i} + \mathbf{j}$.

4 (a) Calculate the magnitude of each of the following vectors:
      (i) $5\mathbf{i} + 12\mathbf{j}$ (ii) $5\mathbf{i} + 7\mathbf{j}$ (iii) $5\mathbf{i} - 7\mathbf{j}$ (iv) $-5\mathbf{i} - 8\mathbf{j}$
      (v) $7\mathbf{i} - 10\mathbf{j}$.
   (b) Calculate the angle made with the positive x-axis by each of the five vectors above.

5 Given that $\mathbf{a} = 2\mathbf{i} + 3\mathbf{j}$ find the magnitude of each of the following vectors and the angle each makes with Ox.
   (a) $\mathbf{a}$ (b) $-\mathbf{a}$ (c) $2\mathbf{a}$ (d) $\tfrac{1}{2}\mathbf{a}$ (e) $-5\mathbf{a}$.

6 Given that $\mathbf{a} = \mathbf{i} + \mathbf{j}$ and $\mathbf{b} = 2\mathbf{i} - \mathbf{j}$ find the magnitude of each of the following vectors and the angle each makes with $0x$.
   (a) $\mathbf{a} + \mathbf{b}$ (b) $\mathbf{a} - \mathbf{b}$ (c) $2\mathbf{a} + \mathbf{b}$ (d) $3\mathbf{a} - 2\mathbf{b}$
   (e) $3\mathbf{b} - \mathbf{a}$.

7 Find a unit vector in the direction of each of the following vectors:

(a) $\mathbf{i} + \mathbf{j}$ (b) $\mathbf{i} - \mathbf{j}$ (c) $3\mathbf{i} - 4\mathbf{j}$ (d) $-3\mathbf{i} + 4\mathbf{j}$
(e) $12\mathbf{i} + 16\mathbf{j}$.

8

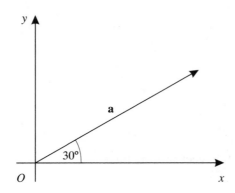

The vector **a** is in the plane Oxy, where Ox, Oy are the coordinate axes, and makes an angle $30°$ with Ox, as in the above diagram. Find the components of **a** along Ox and Oy when a equals:

(a) 2   (b) 3   (c) 6   (d) 10.

9 The vector **a** is in the plane of the coordinate axes Ox and Oy as in question 8, but makes an angle θ with the positive direction Ox. Find the components of **a** along Ox and Oy when $a = 10$ and θ equals:

(a) $60°$ (b) $120°$ (c) $150°$ (d) $240°$ (e) $300°$.

10 Vectors **a** and **b** are in the plane of the coordinate axes Ox and Oy, **a** makes an angle α with Ox, **b** makes an angle β with Ox. Find:

(a) the components of **a** and **b** along Ox, Oy,
(b) the sum of these components along Ox, Oy,
(c) the magnitude of $\mathbf{a} + \mathbf{b}$,

in each of the following cases.

(i) $a = 3,\ b = 4,\ \alpha = 0,\ \beta = 90°$
(ii) $a = 3,\ b = 4,\ \alpha = 0,\ \beta = 60°$
(iii) $a = 3,\ b = 4,\ \alpha = 30°,\ \beta = 60°$
(iv) $a = 5,\ b = 8,\ \alpha = 30°,\ \beta = 60°$
(v) $a = 5,\ b = 8,\ \alpha = 30°,\ \beta = 150°$.

2.4 Using vectors in mechanics

Position vectors

Imagine a particle P moving in a plane. O is a fixed point in the plane. If you know where the point O is, then the position of P is uniquely defined by the vector:

$$\overrightarrow{OP} = \mathbf{r}$$

The vector $\mathbf{r}$ is called the **position vector** of P relative to O.

Example 11

At a given time the cartesian coordinates of the position P of a particle are (2,1). Find the position vector of P relative to O.

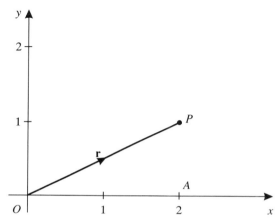

(In chapter 4 of Book P1 the idea of cartesian coordinates is discussed.)

The diagram shows the position of P. Using the $\mathbf{i}$, $\mathbf{j}$ notation:

$$\mathbf{r} = \overrightarrow{OP} = \overrightarrow{OA} + \overrightarrow{AP}$$
$$= 2\mathbf{i} + \mathbf{j}$$

Velocity and acceleration as vectors

If a particle is moving with a *speed* of $5\,\mathrm{m\,s^{-1}}$ then, provided its speed remains constant, it will travel $5\,\mathrm{m}$ in every second. This is true no matter what type of path the particle is moving on. It will travel $10\,\mathrm{m}$ in $2\,\mathrm{s}$, $15\,\mathrm{m}$ in $3\,\mathrm{s}$ and so on. So in this case:

$$\text{distance travelled} = \text{speed} \times \text{time}$$

But if you want a complete picture of what is happening, you also need to know the *direction* in which the particle is moving. The **velocity** of the particle gives us a complete picture.

■ **The velocity of a particle is a vector in the direction of motion whose magnitude is equal to the speed of the particle.**

It is usually denoted by **v**.

(It can be shown that if the position vector of the particle is **r** then $\mathbf{v} = \dfrac{d\mathbf{r}}{dt}$ that is, **v** is the rate of change of **r** with respect to t – see chapter 8 of Book P1.)

When distances are measured in metres and time in seconds, velocities are measured in metres per second $(\mathrm{m\,s^{-1}})$.

If the velocity is a constant vector then:

$$\text{displacement} = \text{velocity} \times \text{time}$$

Example 12

A particle P is moving along the x-axis with a constant speed. At time $t = 0$ P is at the origin. At time $t = 3\,\mathrm{s}$ the particle is at the point $(9,0)$. Find (a) the speed of P, given that the distances are measured in metres. (b) the velocity of P, taking **i** to be the unit vector in the direction of the positive x-axis.

(a) Distance moved in 3 s by P is 9 m.

So: $\qquad$ speed of $P = \dfrac{\text{distance travelled}}{\text{time}} = \dfrac{9}{3} = 3$

The speed of P is $3\,\mathrm{m\,s^{-1}}$.

(b) As the distance moved is along the positive x-axis the velocity of P is $3\mathbf{i}\,\mathrm{m\,s^{-1}}$.

Example 13

A particle P is at the origin O at time $t = 0$. The particle moves with constant velocity and passes through the point with position vector $(6\mathbf{i} + 8\mathbf{j})\,\mathrm{m}$, relative to O, at time $t = 2\mathrm{s}$, where **i**, **j** are unit vectors along the positive x- and y-axes respectively. Find the velocity of P.

From the information given, you see that the displacement of P in 2 s is $(6\mathbf{i} + 8\mathbf{j})\,\mathrm{m}$.

So: $\qquad$ velocity $= \dfrac{\text{displacement}}{\text{time}} = \dfrac{6\mathbf{i} + 8\mathbf{j}}{2}$

The velocity of P is $(3\mathbf{i} + 4\mathbf{j})\,\mathrm{m\,s^{-1}}$.

Everybody has some idea of acceleration from their experience of travelling in buses or cars. Just as velocity tells you how the position of a particle changes with time, the **acceleration** tells you how the velocity changes with time. Since velocity has magnitude and direction so also has acceleration. It is, therefore, a vector. It is usually denoted by **a**. (It can be shown that $\mathbf{a} = \dfrac{d\mathbf{v}}{dt}$.)

■ **Acceleration is the rate of change of velocity with respect to time.**

Hence: $\text{acceleration} = \dfrac{\text{change in velocity}}{\text{time}}$

When the acceleratioon is constant.

Since velocities are measured in metres per second $(m\,s^{-1})$ the acceleration is measured in metres per second per second $(m\,s^{-2})$.

Example 14

A particle P has velocity $(3\mathbf{i} + 2\mathbf{j})\,m\,s^{-1}$ when $t = 0$ and velocity $(7\mathbf{i} + 4\mathbf{j})\,m\,s^{-1}$ at time $t = 2\,s$. The acceleration of P is constant. Find the acceleration.

The change in velocity of P is:

$$(7\mathbf{i} + 4\mathbf{j}) - (3\mathbf{i} + 2\mathbf{j}) = (7\mathbf{i} - 3\mathbf{i}) + (4\mathbf{j} - 2\mathbf{j})$$
$$= (4\mathbf{i} + 2\mathbf{j})\,m\,s^{-1}$$

This takes place in 2 seconds.

$$\text{So acceleration} = \frac{4\mathbf{i} + 2\mathbf{j}}{2}$$

The acceleration of P is $(2\mathbf{i} + \mathbf{j})\,m\,s^{-2}$.

Example 15

A particle P is moving with a constant velocity $(12\mathbf{i} + 5\mathbf{j})\,m\,s^{-1}$. It passes through the point A whose position vector is $(4\mathbf{i} + 5\mathbf{j})\,m$ at $t = 0$. Find:

(a) the speed of the particle
(b) the distance of P from O when $t = 3\,s$.

(a) The speed of the particle is the magnitude of **v**, where $\mathbf{v} = (12\mathbf{i} + 5\mathbf{j})\,m\,s^{-1}$. So:

$$|\mathbf{v}| = \sqrt{(12^2 + 5^2)} = \sqrt{169} = 13$$

The speed of P is $13\,m\,s^{-1}$.

(b) Suppose when $t = 3\,s$ the particle is at the point B as shown in the diagram.

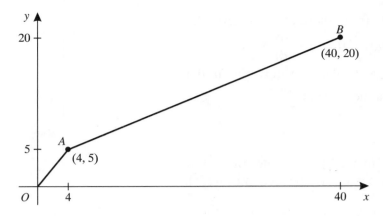

Using: displacement = velocity × time

Gives: $\overrightarrow{AB} = (12\mathbf{i} + 5\mathbf{j}) \times 3$

$$= 36\mathbf{i} + 15\mathbf{j}$$

The position vector of B relative to O is:

$$\overrightarrow{OB} = \overrightarrow{OA} + \overrightarrow{AB}$$
$$= (4\mathbf{i} + 5\mathbf{j}) + (36\mathbf{i} + 15\mathbf{j})$$
$$= 40\mathbf{i} + 20\mathbf{j}$$

So the distance of B from O is:

$$|\overrightarrow{OB}| = \sqrt{(40^2 + 20^2)}$$
$$= \sqrt{2000}$$
$$= 44.7\,\text{m}$$

Exercise 2D

In these questions, $\mathbf{i}$ and $\mathbf{j}$ are unit vectors in the direction of the positive x- and y-axes respectively. The units of measure are metres and seconds.

1 Points A, B, C, D have coordinates $(2,3)$, $(3,-1)$, $(-1,-1)$ and $(0,-4)$ respectively. Plot these points on coordinate axes Ox and Oy and write down in terms of $\mathbf{i}$ and $\mathbf{j}$ the position vector of:

    (a) A relative to the origin O

    (b) D relative to the origin O

    (c) A relative to D

    (d) B relative to D

    (e) C relative to D.

2 At time $t = 0$ a particle is at the point P, position vector **r**, with velocity vector **v**. Plot the point P on a diagram and draw the velocity vector to show the direction in which the particle is moving in each of the following cases.

(a) **r** = 2**i**, **v** = 2**j**

(b) **r** = 2**i**, **v** = −**j**

(c) **r** = 2**j**, **v** = **i** + **j**

(d) **r** = −3**i**, **v** = 3**i** + **j**

(e) **r** = **i** + **j**, **v** = **i** − **j**.

3 For each part of question 2 find the position vector of the particle when $t = 1$, given that the velocity has remained constant between $t = 0$ and $t = 1$.

4 A particle P moving with constant velocity **v** has position vector **a** initially and **b** at time t. Find **v** when:

(a) **a** = 2**i**, **b** = 4**i** + 4**j**, $t = 2$

(b) **a** = 2**i**, **b** = 4**j**, $t = 2$

(c) **a** = −2**i**, **b** = 2**i** + 8**j**, $t = 4$

(d) **a** = −**i** + 3**j**, **b** = 4**i** −2**j**, $t = 5$

(e) **a** = −**i** −4**j**, **b** = 2**i** + 2**j**, $t = 3$.

5 A particle has position vector **a** initially and position vector **b** 2 seconds later. Find the velocity vector and the speed of the particle when:

(a) **a** = 3**i**, **b** = 9**i**

(b) **a** = 3**i**, **b** = 9**i** + 8**j**

(c) **a** = 3**i**, **b** = −3**i** + 8**j**

(d) **a** = −3**i**, **b** = 3**i** + 8**j**

(e) **a** = −3**i**, **b** = 5**i** + 6**j**.

6 A particle P has velocity **u** initially and acceleration **a**. Find the velocity and the speed after t seconds when:

(a) **u** = 2**i**, **a** = **i**, $t = 3$

(b) **u** = 3**i**, **a** = **j**, $t = 4$

(c) **u** = 5**i**, **a** = 3**j**, $t = 4$

(d) **u** = 5**i**, **a** = −3**j**, $t = 4$

(e) **u** = 8**i**, **a** = 3**j**, $t = 5$.

7 A particle P has velocity **u** initially and **v** after t seconds. Find its acceleration, assumed constant, when:

(a) **u** = 2**i**, **v** = 4**i**, $t = 4$

(b) $\mathbf{u} = 2\mathbf{i}$, $\mathbf{v} = 4\mathbf{j}$, $t = 2$

(c) $\mathbf{u} = 2\mathbf{i} + 3\mathbf{j}$, $\mathbf{v} = -3\mathbf{i} - 2\mathbf{j}$, $t = 5$

(d) $\mathbf{u} = \mathbf{i} + \mathbf{j}$, $\mathbf{v} = -3\mathbf{i} + 2\mathbf{j}$, $t = 1$

(e) $\mathbf{u} = \mathbf{i} + 2\mathbf{j}$, $\mathbf{v} = 4\mathbf{i} - 3\mathbf{j}$, $t = 1$.

8 A particle has velocity $2\mathbf{i} + 3\mathbf{j}$ initially and velocity $5\mathbf{i} - \mathbf{j}$ 5 seconds later. Find its acceleration, assumed constant, and the magnitude of that acceleration.

9 A particle has position vector $2\mathbf{i} + \mathbf{j}$ initially and is moving with speed $10\,\mathrm{m\,s}^{-1}$ in the direction $3\mathbf{i} - 4\mathbf{j}$. Find its position vector when $t = 3$ and the distance it has travelled in those 3 seconds.

10 A particle has an initial velocity of $\mathbf{i} + 2\mathbf{j}$ and is accelerating in the direction $\mathbf{i} + \mathbf{j}$. If the magnitude of the acceleration is $5\sqrt{2}$, find the velocity vector and the speed of the particle after 2 seconds.

2.5 Differentiating and integrating vectors

This section requires a basic knowledge of differentiation and integration. If possible study chapters 8 and 9 of Book P1 before you study this section.

Differentiating vectors

In mechanics you often need to know how displacements, velocities and accelerations change with time. As you will see in section 3.4, in general it is necessary to use differentiation when considering these changes.

As displacements, velocities and accelerations are all vectors, you first have to look at how to differentiate vectors with respect to time.

Use a dot as a quick way to denote differentiation with respect to time. With this notation:

$$\frac{\mathrm{d}\mathbf{r}}{\mathrm{d}t} = \dot{\mathbf{r}} \quad \text{and} \quad \frac{\mathrm{d}^2\mathbf{r}}{\mathrm{d}t^2} = \ddot{\mathbf{r}}$$

(The number of dots denotes the number of differentiations to be carried out.)

The basic rules of differentiation developed in Book P1, chapter 8, hold for vectors. So:

$$\frac{d}{dt}(\mathbf{r}_1 + \mathbf{r}_2) = \frac{d\mathbf{r}_1}{dt} + \frac{d\mathbf{r}_2}{dt}$$

and if $\mathbf{r} = f(t)\mathbf{c}$, where $\mathbf{c}$ is a constant vector, then:

$$\frac{d\mathbf{r}}{dt} = \frac{d}{dt}[f(t)\mathbf{c}] = \frac{df}{dt}\mathbf{c}$$

The unit vectors $\mathbf{i}$ and $\mathbf{j}$ are constant vectors, so if $\mathbf{r}$ is given in the $\mathbf{i}, \mathbf{j}$ notation then:

$$\mathbf{r} = x\mathbf{i} + y\mathbf{j}$$
$$\dot{\mathbf{r}} = \dot{x}\mathbf{i} + \dot{y}\mathbf{j}$$
$$\ddot{\mathbf{r}} = \ddot{x}\mathbf{i} + \ddot{y}\mathbf{j}$$

Example 16

The position vector of a particle P at time t s is $\mathbf{r} = (2t - 1)\mathbf{i} + t^2\mathbf{j}$, where $\mathbf{r}$ is measured in metres. Find the initial position vector and velocity of P and show that the acceleration of P is constant.

When $t = 0$: $\qquad\qquad\qquad \mathbf{r} = -\mathbf{i}$

Therefore the initial position vector of P is $-\mathbf{i}$ m.

Differentiating with respect to t gives:

$$\dot{\mathbf{r}} = 2\mathbf{i} + 2t\mathbf{j}$$

So when $t = 0$: $\qquad\qquad \dot{\mathbf{r}} = 2\mathbf{i}\,\mathrm{m\,s}^{-1}$

Differentiating again gives:

$$\ddot{\mathbf{r}} = 2\mathbf{j}\,\mathrm{m\,s}^{-2}$$

So the acceleration is a constant vector.

Integrating vectors

As you will see in chapter 3, you also need to be able to integrate vectors. In particular you should be able to obtain the velocity $\mathbf{v}$ when you are given the acceleration $\mathbf{a}$ and also to find the position vector $\mathbf{r}$ when you are given the velocity $\mathbf{v}$. The integration of vectors written in the $\mathbf{i}, \mathbf{j}$ notation follows the rules of integration given in Book P1, chapter 9.

If: $$\ddot{\mathbf{r}} = f(t)\mathbf{i} + g(t)\mathbf{j}$$

then: $$\dot{r} = \int \ddot{r}\, dt$$

$$= \mathbf{i}\int f(t)\, dt + \mathbf{j}\int g(t)\, dt + C_1\mathbf{i} + C_2\mathbf{j}$$

■ **When you integrate a vector you must add an arbitrary constant vector $C_1\mathbf{i} + C_2\mathbf{j}$, where C_1 and C_2 are constants.**

This is just a generalisation of the idea of adding an arbitrary constant when you integrate a function.

Similarly, if: $$\dot{\mathbf{r}} = F(t)\mathbf{i} + G(t)\mathbf{j}$$

then: $$\mathbf{r} = \mathbf{i}\int F(t)\, dt + \mathbf{j}\int G(t)\, dt + k_1\mathbf{i} + k_2\mathbf{j}$$

where k_1 and k_2 are constants.

Example 17

A particle P moves so that at time t:

$$\dot{\mathbf{r}} = 2t\mathbf{i} + 3t^2\mathbf{j}$$

When $t = 0$ its position vector is $4\mathbf{i} - 3\mathbf{j}$. Find the position vector $\mathbf{r}$ of P at time t.

As: $$\dot{\mathbf{r}} = 2t\mathbf{i} + 3t^2\mathbf{j}$$

then: $$\mathbf{r} = \mathbf{i}\int 2t\, dt + \mathbf{j}\int 3t^2\, dt + k_1\mathbf{i} + k_2\mathbf{j}$$

$$= t^2\mathbf{i} + t^3\mathbf{j} + k_1\mathbf{i} + k_2\mathbf{j}$$

At $t = 0$ this gives:

$$\mathbf{r} = 0\mathbf{i} + 0\mathbf{j} + k_1\mathbf{i} + k_2\mathbf{j}$$

But this must be equal to $4\mathbf{i} - 3\mathbf{j}$.

Equating $\mathbf{i}$ parts gives: $\qquad\qquad k_1 = 4$

Equating $\mathbf{j}$ parts gives: $\qquad\qquad k_2 = -3$

So: $$\mathbf{r} = t^2\mathbf{i} + t^3\mathbf{j} + 4\mathbf{i} - 3\mathbf{j}$$

$$= (t^2 + 4)\mathbf{i} + (t^3 - 3)\mathbf{j}$$

The position vector of P at time t is $\mathbf{r} = (t^2 + 4)\mathbf{i} + (t^3 - 3)\mathbf{j}$.

Example 18

A particle P moves so that $\ddot{\mathbf{r}} = p\mathbf{i} + q\mathbf{j}$ where p and q are constants. (This means $\ddot{\mathbf{r}}$ is a constant vector.) When $t = 0$, $\dot{\mathbf{r}} = \mathbf{0}$ and when $t = 1$, $\dot{\mathbf{r}} = 3\mathbf{i} - 2\mathbf{j}$. Find:

(a) $\dot{\mathbf{r}}$ at time t
(b) $|\dot{\mathbf{r}}|$ when $t = 3$.

(a) You know that: $\quad\quad\quad\quad \ddot{\mathbf{r}} = p\mathbf{i} + q\mathbf{j}$

So: $\quad\quad\quad\quad\quad\quad\quad\quad \dot{\mathbf{r}} = \mathbf{i}\int p\,dt + \mathbf{j}\int q\,dt + C_1\mathbf{i} + C_2\mathbf{j}$

$$= pt\mathbf{i} + qt\mathbf{j} + C_1\mathbf{i} + C_2\mathbf{j}$$

Since: $\dot{\mathbf{r}} = \mathbf{0}$ when $t = 0$:

$$\mathbf{0} = 0\mathbf{i} + 0\mathbf{j} + C_1\mathbf{i} + C_2\mathbf{j}$$

and so: $\quad\quad\quad\quad\quad\quad C_1 = 0 \text{ and } C_2 = 0$

So: $\quad\quad\quad\quad\quad\quad\quad\quad \dot{\mathbf{r}} = pt\mathbf{i} + qt\mathbf{j}$

When $t = 1$ you know that $\dot{\mathbf{r}} = 3\mathbf{i} - 2\mathbf{j}$.

So: $\quad\quad\quad\quad\quad\quad\quad\quad p\mathbf{i} + q\mathbf{j} = 3\mathbf{i} - 2\mathbf{j}$

$$p = 3$$

$$q = -2$$

Hence: $\quad\quad\quad\quad\quad\quad\quad \dot{\mathbf{r}} = 3t\mathbf{i} - 2t\mathbf{j}$

(b) When $t = 3$: $\quad\quad\quad\quad \dot{\mathbf{r}} = 3(3)\mathbf{i} - 2(3)\mathbf{j}$

$$= 9\mathbf{i} - 6\mathbf{j}$$

and $\quad\quad\quad\quad\quad\quad\quad |\dot{\mathbf{r}}| = \sqrt{(9^2 + 6^2)}$

$$= \sqrt{(81 + 36)}$$

$$= \sqrt{117} = 10.8$$

Example 19

At time $t = 0$ a particle P is at the point with position vector $(30\mathbf{i} + 35\mathbf{j})$ m relative to an origin O. P moves so that $\ddot{\mathbf{r}}$ is constant and equal to $(3\mathbf{i} + 4\mathbf{j})\,\mathrm{m\,s^{-2}}$. When $t = 0$, $\dot{\mathbf{r}} = \mathbf{0}$. Find:

(a) the vector $\dot{\mathbf{r}}$ when $t = 4$
(b) the distance of P from O at this time.

(a) As $\ddot{\mathbf{r}} = 3\mathbf{i} + 4\mathbf{j}$, integrating with respect to t gives:

$$\dot{\mathbf{r}} = \int \ddot{\mathbf{r}}\,dt = \mathbf{i}\int 3\,dt + \mathbf{j}\int 4\,dt + C_1\mathbf{i} + C_2\mathbf{j}$$

$$= 3t\mathbf{i} + 4t\mathbf{j} + C_1\mathbf{i} + C_2\mathbf{j}$$

Using $\dot{\mathbf{r}} = \mathbf{0}$ when $t = 0$ gives:

$$\mathbf{0} = C_1\mathbf{i} + C_2\mathbf{j}$$

and so: $\quad\quad\quad\quad\quad\quad C_1 = 0 \text{ and } C_2 = 0$

Hence: $\quad\quad\quad\quad\quad\quad\quad \dot{\mathbf{r}} = 3t\mathbf{i} + 4t\mathbf{j}$

Substituting $t = 4$ in this result gives $\dot{\mathbf{r}}$ when $t = 4$ as:

$$\dot{\mathbf{r}} = (12\mathbf{i} + 16\mathbf{j})\,\mathrm{m\,s^{-1}}$$

(b) The distance of P from O is just the magnitude of $\mathbf{r}$. So you must first find $\mathbf{r}$.

As $\dot{\mathbf{r}} = 3t\mathbf{i} + 4t\mathbf{j}$, integrating with respect to t gives:

$$\mathbf{r} = \int \dot{\mathbf{r}}\,dt$$
$$= \mathbf{i}\int 3t\,dt + \mathbf{j}\int 4t\,dt + k_1\mathbf{i} + k_2\mathbf{j}$$
$$= \mathbf{i}\frac{3t^2}{2} + \mathbf{j}\frac{4t^2}{2} + k_1\mathbf{i} + k_2\mathbf{j}$$

When $t = 0$, $\mathbf{r} = 30\mathbf{i} + 35\mathbf{j}$, and so:

$$30\mathbf{i} + 35\mathbf{j} = 0\mathbf{i} + 0\mathbf{j} + k_1\mathbf{i} + k_2\mathbf{j}$$

Equating $\mathbf{i}$ and $\mathbf{j}$ components gives:

$$k_1 = 30, k_2 = 35$$

So:
$$\mathbf{r} = \frac{3t^2}{2}\mathbf{i} + 2t^2\mathbf{j} + 30\mathbf{i} + 35\mathbf{j}$$

When $t = 4$:
$$\mathbf{r} = 24\mathbf{i} + 32\mathbf{j} + 30\mathbf{i} + 35\mathbf{j}$$
$$= 54\mathbf{i} + 67\mathbf{j}$$

The magnitude of $\mathbf{r}$ is:

$$|\mathbf{r}| = \sqrt{(54^2 + 67^2)}$$
$$= 86.1$$

So the distance of P from O at this time is $86.1\,\mathrm{m}$.

Exercise 2E

In these questions, $\mathbf{i}$ and $\mathbf{j}$ are unit vectors in the direction of the positive x- and y-axes, respectively. The units of measure are metres and seconds.

1 The position of a particle P at time t is determined by its position vector $\mathbf{r}$, from which the velocity vector can be found. The diagram shows the position of P and the direction of its velocity when $\mathbf{r} = t\mathbf{i} + t^2\mathbf{j}$, and $t = 1$.

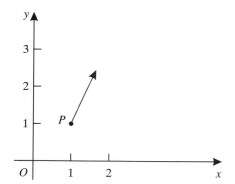

Draw similar diagrams showing the position and direction of motion of the particle when $t = 0$, $t = 1$ and $t = 2$ in each of the following cases:

(a) $\mathbf{r} = 2\mathbf{i} + t\mathbf{j}$

(b) $\mathbf{r} = 2\mathbf{i} + t^2\mathbf{j}$

(c) $\mathbf{r} = t\mathbf{i} + t\mathbf{j}$

(d) $\mathbf{r} = t\mathbf{i} + 2\mathbf{j}$

(e) $\mathbf{r} = t^2\mathbf{i} + \frac{1}{2}t^3\mathbf{j}$.

2 Find the velocity vector $\mathbf{v}$ for each part of question 1, and the velocity and speed in each case when $t = 1$.

3 A particle P has position vector $\mathbf{r}$. Find $\dot{\mathbf{r}}$ and $\ddot{\mathbf{r}}$ in each of the following cases:

(a) $\mathbf{r} = t\mathbf{i} + t^2\mathbf{j}$

(b) $\mathbf{r} = \frac{1}{2}t^2\mathbf{i} + t\mathbf{j}$

(c) $\mathbf{r} = t^3\mathbf{i} + t\mathbf{j}$

(d) $\mathbf{r} = t^3\mathbf{i} + \mathbf{j}$

(e) $\mathbf{r} = \frac{1}{3}t^3\mathbf{i} + \frac{1}{4}t^4\mathbf{j}$.

4 A particle P has position vector $\mathbf{r}$. Find the velocity and acceleration vectors in each of the following cases:

(a) $\mathbf{r} = t^2\mathbf{i} + \mathbf{j}$

(b) $\mathbf{r} = t^2\mathbf{i} + t\mathbf{j}$

(c) $\mathbf{r} = t^2\mathbf{i} + t^2\mathbf{j}$

(d) $\mathbf{r} = t^3\mathbf{i} + t^3\mathbf{j}$

(e) $\mathbf{r} = \frac{1}{4}t^4\mathbf{i} + \frac{1}{3}t^3\mathbf{j}$.

5 For each part of question 4 find the speed of the particle when $t = 2$.

6 A particle P is initially at the origin and has velocity vector **v**. Find the position vector **r** and the acceleration vector **a** in each of the following cases:

(a) $\mathbf{v} = 2t\mathbf{i} + 2\mathbf{j}$

(b) $\mathbf{v} = 2t\mathbf{i} + 3t^2\mathbf{j}$

(c) $\mathbf{v} = \mathbf{i} + 3t^2\mathbf{j}$

(d) $\mathbf{v} = 4t^3\mathbf{i} + 3t^2\mathbf{j}$

(e) $\mathbf{v} = t^2\mathbf{i} + 12t^3\mathbf{j}$.

7 A particle P is initially at the origin O with velocity **i**. Find the velocity vector **v** and the position vector **r** in each of the following cases, where **a** is the acceleration vector:

(a) $\mathbf{a} = \mathbf{i}$

(b) $\mathbf{a} = \mathbf{j}$

(c) $\mathbf{a} = \mathbf{i} + \mathbf{j}$

(d) $\mathbf{a} = 6t\mathbf{i} + \mathbf{j}$

(e) $\mathbf{a} = \mathbf{i} + 6t\mathbf{j}$.

8 Sketch diagrams similar to those in question 1 to show the path of the particle P for the motion described in question 4(a), (b), (c).

SUMMARY OF KEY POINTS

1 Scalar quantities are completely specified by their magnitude.

2 Vector quantities require both their size (magnitude) and direction to be specified.

3 Vectors are added by using the triangle law of addition:

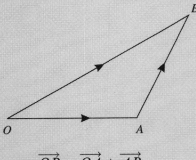

$$\overrightarrow{OB} = \overrightarrow{OA} + \overrightarrow{AB}$$

4

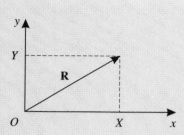

Using the **i**, **j** notation:
$$\mathbf{R} = X\mathbf{i} + Y\mathbf{j}$$

5 The velocity of a particle is the rate of change with respect to time of its displacement.

6 The acceleration of a particle is the rate of change with respect to time of its velocity.

7 If: $\mathbf{r} = x\mathbf{i} + y\mathbf{j}$

then: $\dot{\mathbf{r}} = \dot{x}\mathbf{i} + \dot{y}\mathbf{j}$

and: $\ddot{\mathbf{r}} = \ddot{x}\mathbf{i} + \ddot{y}\mathbf{j}$

Kinematics of a particle

3

What is kinematics?

Kinematics is the study of the motion of a particle – its speed, acceleration, and the path it follows. Kinematics does not deal with the forces causing the motion but only with the description of the motion itself. (The forces that cause motion are discussed in chapter 5.)

3.1 Motion in a straight line with constant acceleration

Imagine a particle moving in a straight line with a constant or uniform acceleration a for a time interval t. During this time its velocity will change from u, its initial velocity, to v, its final velocity, while it travels a distance s. If you know any three of the quantities a, t, u, v and s you can calculate the remaining two.

The particle's velocity has changed from u to v in time t. Its acceleration is:

$$\text{acceleration} = \frac{\text{change in velocity}}{\text{change in time}}$$

Or:
$$a = \frac{v - u}{t}$$

So:
$$at = v - u$$

■
$$v = u + at \qquad (1)$$

As the acceleration is constant, the average velocity is the average of the initial and final velocities:

$$\text{average velocity} = \frac{u + v}{2}$$

But the average velocity is also:

$$\text{average velocity} = \frac{\text{displacement}}{\text{time}} = \frac{s}{t}$$

So :

$$\frac{s}{t} = \frac{u+v}{2}$$

■
$$s = \left(\frac{u+v}{2}\right)t \qquad (2)$$

From equation (1), $v = u + at$. Substituting this expression for v into equation (2) gives:

$$s = \left(\frac{u+u+at}{2}\right)t$$

■
$$s = ut + \tfrac{1}{2}at^2 \qquad (3)$$

Making t the subject of equation (1) gives:

$$t = \frac{v-u}{a}$$

and substituting this into equation (2) gives:

$$s = \left(\frac{u+v}{2}\right)\left(\frac{v-u}{a}\right)$$
$$2as = uv - u^2 + v^2 - uv$$
$$2as = v^2 - u^2$$
$$v^2 = u^2 + 2as \qquad (4)$$

■

Equations (1) – (4) are very important. You need to remember them, but remember too that they apply to *motion with constant acceleration* only. They are often called the **uniform acceleration equations**.

Example 1

A particle is moving in a straight line from O to A with a constant acceleration of 2 m s^{-2}. Its velocity at A is 30 m s^{-1} and it takes 15 seconds to travel from O to A. Find (a) the particle's velocity at O and (b) the distance OA.

When solving problems of this type it can be useful to start by listing the known quantities. Known quantities are:

$$a = 2 \text{ m s}^{-2}$$
$$v = 30 \text{ m s}^{-1}$$
$$t = 15 \text{ s}$$

(a) To find u, the particle's initial velocity at O, use equation (1),
$v = u + at$.
Substituting the known quantities gives:
$$30 = u + 2 \times 15$$
So:
$$u = 0$$

The velocity at O is 0 m s^{-1}

(b) To find the distance OA use equation (3), $s = ut + \frac{1}{2}at^2$.
$$s = 0 + \frac{1}{2} \times 2 \times 15^2$$
So:
$$s = 225$$

The distance OA is 225 m.

Example 2

A train starts from rest at a station S and moves with constant acceleration. It passes a signal box B 15 seconds later with a speed of 81 km h^{-1}. Modelling the train as a particle find the acceleration of the train in m s^{-2} and the distance in metres between the station and the signal box.

In this example metres, kilometres, hours and seconds are all present. For the uniform acceleration equations to be valid, consistent units must be used. It is usual to work with metres and seconds, so the first step is to change 81 km h^{-1} to m s^{-1}.

As 1 km $=$ 1000 m and 1 hour $=$ 60 $\times$ 60 seconds:

$$81 \, \text{km} \, \text{h}^{-1} = \frac{81 \times 1000}{60 \times 60} \, \text{m} \, \text{s}^{-1}$$
$$= 22.5 \, \text{m} \, \text{s}^{-1}$$

The uniform acceleration equations can now be used.

Known quantities are:
$$u = 0 \, \text{m} \, \text{s}^{-1}$$
$$v = 22.5 \, \text{m} \, \text{s}^{-1}$$
$$t = 15 \, \text{s}$$

As:
$$v = u + at$$
$$22.5 = 0 + a \times 15$$
So:
$$a = \frac{22.5}{15} = 1.5$$

The acceleration of the train is 1.5 m s^{-2}.

To find s use equation (2), $s = \left(\dfrac{u + v}{2}\right)t$.

Substituting the known quantities gives:

$$s = \left(\frac{0 + 22.5}{2}\right) \times 15$$

So: $\qquad\qquad\qquad s = 168.75$

The distance between the station and the signal box is 169 m.

Retardation

If a particle is slowing down its initial velocity is greater than its final velocity. So its acceleration $\dfrac{v - u}{t}$ is negative. A negative acceleration is called a **retardation**. To represent a retardation in the uniform acceleration equations you need to use a negative value for the acceleration a, such as $-2\,\mathrm{m\,s}^{-2}$.

Example 3

A boy on a skateboard is travelling up a hill. He experiences a constant retardation of magnitude $2\,\mathrm{m\,s}^{-2}$. Given that his speed at the bottom of the hill was $10\,\mathrm{m\,s}^{-1}$ determine how far he will travel before he comes to rest.

A retardation of $2\,\mathrm{m\,s}^{-2}$ is an acceleration of $-2\,\mathrm{m\,s}^{-2}$.

Known quantities are:
$$a = -2\,\mathrm{m\,s}^{-2}$$
$$u = 10\,\mathrm{m\,s}^{-1}$$
$$v = 0\,\mathrm{m\,s}^{-1}$$

Using: $\qquad\qquad v^2 = u^2 + 2as$

Gives: $\qquad\qquad 0 = 10^2 + 2 \times (-2)s$

$$s = \frac{100}{4} = 25$$

The boy travels a distance of 25 m before coming to rest.

Exercise 3A

1   A particle moves in a straight line with uniform acceleration $5\,\mathrm{m\,s}^{-2}$. It starts from rest when $t = 0$. Find its velocity when $t = 3\,\mathrm{s}$.

2 A particle moves in a straight line. When $t = 0$ its velocity is $3\,\text{m}\,\text{s}^{-1}$. When $t = 4$ its velocity is $12\,\text{m}\,\text{s}^{-1}$. Find its acceleration, assumed to be constant.

3 A particle moves in a straight line with constant retardation $4\,\text{m}\,\text{s}^{-2}$. When $t = 3\,\text{s}$ its velocity is $5\,\text{m}\,\text{s}^{-1}$. Find its initial velocity.

4 A particle moves in a straight line with constant acceleration $5\,\text{m}\,\text{s}^{-2}$. How long will it take for the particle's velocity to increase from $2\,\text{m}\,\text{s}^{-1}$ to $24\,\text{m}\,\text{s}^{-1}$?

5 A particle starts from rest and moves in a straight line with constant acceleration $4\,\text{m}\,\text{s}^{-2}$ for $3\,\text{s}$. How far does it travel?

6 A particle moves along a line PQ with constant acceleration $3\,\text{m}\,\text{s}^{-2}$. If PQ is $2\,\text{m}$ and the particle takes $0.5\,\text{s}$ to travel from P to Q, what was its velocity at P?

7 A particle travels in a straight line with uniform acceleration $8\,\text{m}\,\text{s}^{-2}$. If its initial velocity is $2\,\text{m}\,\text{s}^{-1}$ how long will it take to travel $6\,\text{m}$?

8 A particle moving in a straight line experiences a constant retardation of $6\,\text{m}\,\text{s}^{-2}$. How long will it take to decrease its speed from $20\,\text{m}\,\text{s}^{-1}$ to $8\,\text{m}\,\text{s}^{-1}$ and how far will it travel while doing so?

9 A particle is moving with uniform acceleration. If it starts from rest and $5\,\text{s}$ later has a speed of $18\,\text{m}\,\text{s}^{-1}$ find the distance it has travelled.

10 A particle is moving in a straight line with uniform acceleration. If it travels $120\,\text{m}$ while increasing speed from $5\,\text{m}\,\text{s}^{-1}$ to $25\,\text{m}\,\text{s}^{-1}$ find its acceleration.

11 A car is accelerating uniformly while travelling along a straight road. Its speed increases from $8\,\text{m}\,\text{s}^{-1}$ to $22\,\text{m}\,\text{s}^{-1}$ in $10\,\text{s}$. Modelling the car as a particle find the distance travelled during this time and the acceleration of the car.

12 A car is travelling along a straight lane at $20\,\text{m}\,\text{s}^{-1}$ when the driver sees a tractor blocking the lane $30\,\text{m}$ ahead. The car's brakes can produce a retardation of $5\,\text{m}\,\text{s}^{-2}$. With what speed does the car hit the tractor?

13 The brakes of a train can produce a retardation of $1.7\,\text{m}\,\text{s}^{-2}$. If the train is travelling at $100\,\text{km}\,\text{h}^{-1}$ and applies its brakes what distance does it travel before stopping? If the driver applies the brakes 15 m too late to stop at a station with what speed is the train travelling when it passes through that station?

14 A particle is moving along a straight line. It passes points A, B and C on the line at $t = 0$, $t = 3$ s and $t = 6$ s respectively. If AC is 60 m and the velocity of the particle at A is 4 m s^{-1} find the acceleration of the particle (assumed uniform) and the distance AB.

15 A, B and C are three points on a straight road such that $AB = 80$ m and $BC = 60$ m. A car travelling with uniform acceleration passes A, B and C at times $t = 0$, $t = 4$ s and $t = 6$ s respectively. Modelling the car as a particle find its acceleration and its velocity at A.

3.2 Vertical motion under gravity

If you ignore the effect of air resistance on a particle which is moving vertically you can obtain a simplified mathematical model for the particle's motion. This model is, however, a good approximation to the real-life situation.

The particle is moving with a constant acceleration g due to gravity. This acceleration has a numerical value of approximately 9.8 m s^{-2}. Throughout this book, whenever a numerical value of g is required it will be taken to be 9.8 m s^{-2}.

As the acceleration is constant, the uniform acceleration equations can be used.

When solving a problem, you must choose a direction, either upwards or downwards, to be positive. An arrow indicating the positive direction in your solution will help to remind you.

Example 4
A marble falls off a shelf which is 1.6 m above the floor. Find:
(a) the time it takes to reach the floor
(b) the speed with which it will reach the floor.

In this question the motion is always downwards so choose downwards to be the positive direction. The constant acceleration is in this direction. Since the particle falls off a shelf it will be considered to have an initial speed of $0 \, \text{m s}^{-1}$.

Known quantities are:

$$u = 0 \, \text{m s}^{-1}$$

$$\downarrow \qquad s = 1.6 \, \text{m}$$

$$+\text{ve} \qquad a = 9.8 \, \text{m s}^{-1}$$

(a) Use $\qquad s = ut + \frac{1}{2}at^2$ to find t.

Substituting the known quantities gives:

$$1.6 = 0 + \frac{1}{2} \times 9.8t^2$$

$$t^2 = \frac{1.6 \times 2}{9.8}$$

$$t = 0.5714$$

The time taken to reach the floor is $0.571 \, \text{s}$.

(b) Using: $\qquad v^2 = u^2 + 2as$

Gives: $\qquad v^2 = 0 + 2 \times 9.8 \times 1.6$

$$v = 5.6$$

The marble reaches the floor with a speed of $5.6 \, \text{m s}^{-1}$.

■ **If a problem involves upward motion, remember the acceleration is always downwards.**

Example 5

A particle P is projected vertically upwards from a point O with a speed of $28 \, \text{m s}^{-1}$. Find:
(a) the greatest height h above O reached by P
(b) the total time before P returns to O
(c) the total distance travelled by the particle.

In this question the particle is initially moving upwards and the distance required is measured upwards, so take upwards as the positive direction.

(a) Consider the motion to the highest point:

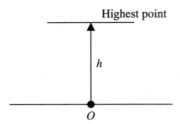

At the highest point the particle is no longer moving upwards but has not yet started to move downwards so its velocity is $0\,\mathrm{m\,s^{-1}}$.

Known quantities are:

$$u = 28\,\mathrm{m\,s^{-1}}$$

+ve $\quad v = 0\,\mathrm{m\,s^{-1}}$

$\uparrow \quad\quad a = 9.8\,\mathrm{m\,s^{-2}}$

$$s = h\,\mathrm{m}$$

Using: $\quad\quad\quad\quad v^2 = u^2 + 2as$

Gives: $\quad\quad\quad\quad 0 = 28^2 + 2 \times (-9.8)h$

$$h = \frac{28^2}{2 \times 9.8} = 40$$

The greatest height above O reached by the particle is 40 m.

(b) When the particle returns to O its displacement from O is zero.

Known quantities are: $\quad s = 0\,\mathrm{m}$

+ ve $\quad u = 28\,\mathrm{m\,s^{-1}}$

$\uparrow \quad\quad a = -9.8\,\mathrm{m\,s^{-2}}$

Using: $\quad\quad\quad\quad s = ut + \frac{1}{2}at^2 :$

Gives: $\quad\quad\quad\quad 0 = 28\,t + \frac{1}{2} \times (-9.8)t^2$

$$0 = 28\,t - 4.9\,t^2$$

Factorising this quadratic equation as shown is Book P1, chapter 2, gives:

$$0 = (28 - 4.9\,t)t$$

So: $\quad\quad\quad\quad t = 0 \quad$ (at start of motion)

Or: $\quad\quad\quad\quad t = \frac{28}{4.9} = 5.714$

So the particle returns to its point of projection in 5.71 s.

(c) The total distance travelled by the particle is the distance it travels upwards plus the distance it travels downwards. This is twice the greatest height. The total distance travelled is 40 m.

Example 6

A stone is catapulted vertically upwards with a velocity of $24.5\,\mathrm{m\,s^{-1}}$. Modelling the stone as a particle moving under gravity alone, find for how long its height exceeds 29.4 m.

Known quantities are: $u = 24.5\,\mathrm{m\,s^{-1}}$

$+\,\mathrm{ve}$ $s = 29.4\,\mathrm{m}$

$\uparrow$ $a = -9.8\,\mathrm{m\,s^{-2}}$

Using: $$s = ut + \tfrac{1}{2}at^2$$

Gives: $$29.4 = 24.5\,t - \tfrac{1}{2}\times 9.8\,t^2$$

$$4.9\,t^2 - 24.5\,t + 29.4 = 0$$

$\div\,4.9$: $$t^2 - 5t + 6 = 0$$

Factorising, as shown in Book P1 chapter 2, gives:

$$(t-3)(t-2) = 0$$

$$t = 2 \text{ or } t = 3$$

The stone is at a height of 29.4 m on its way up at 2 s after projection and again at 3 s, as it comes down.

Its height is greater than 29.4 m for $(3-2)\mathrm{s} = 1\,\mathrm{s}$.

Exercise 3B

Whenever a numerical value of g is required take $g = 9.8\,\mathrm{m\,s^{-2}}$.

1 A ball falls off a table which is 1 m high. Find the speed with which the ball hits the floor.

2 A stone is dropped from the top of a tower which is 20 m high. How long does it take to reach the ground?

3 A book is dropped from the top of a tower. It hits the ground with a speed of $15\,\mathrm{m\,s^{-1}}$. By modelling the book as a particle find the height of the tower.

4 A ball is thrown vertically upwards with a speed of $10\,\mathrm{m\,s^{-1}}$. Find (a) the highest point it reaches (b) the total time it is in the air.

5 A stone is catapulted vertically upwards with a velocity of $25\,\mathrm{m\,s^{-1}}$ from a point 2 m above the ground. Find (a) its velocity when it hits the ground (b) the time it takes to reach the ground.

6 Water from a fountain rises to a height of 6 m. By modelling the drops of water as particles find the speed of the water as it leaves the nozzle.

7 A ball is thrown vertically upwards with a speed of $29\,\mathrm{m\,s^{-1}}$. It hits the ground 6 seconds later. By modelling the ball as a particle find the height above the ground from which it was thrown.

8 A ball is thrown vertically upwards with a speed of $15\,\mathrm{m\,s^{-1}}$ from a point 1 m above the ground. Find the speed with which it hits the floor. If it rebounds with a speed which is half the speed with which it hits the floor, find its greatest height after the first bounce.

9 A particle is projected upwards with a speed of $14\,\mathrm{m\,s^{-1}}$. Find for how long it is above 2 m.

10 A stone is dropped from the top of a tower. One second later another stone is thrown vertically downwards from the same point with a velocity of $14\,\mathrm{m\,s^{-1}}$. If they hit the ground together find the height of the tower.

11 A book falls from a window which is 30 m above the ground. Model the book as a particle and hence find the time taken to reach the ground. What assumptions have you made when adopting this model? What effect is this likely to have had on your answer?

3.3 Speed-time graphs

If a particle is moving with a constant acceleration which causes its speed to change from u to v in time t, a graph showing the particle's change in speed will be a straight line. For speed-time graphs, time is always plotted along the horizontal axis.

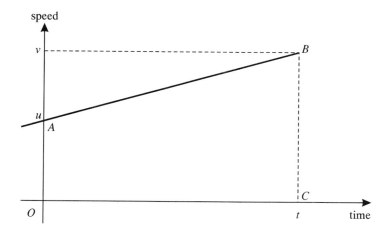

The gradient of the line AB is $\dfrac{v-u}{t}$. This is also the $\dfrac{\text{change in velocity}}{\text{time}}$, that is the acceleration of the particle. Hence the gradient of a speed-time graph gives the acceleration of the particle. $OABC$ is a trapezium with area $\frac{1}{2}(u+v)t$. This is also the distance travelled by the particle as given by the uniform acceleration equations. So the area under the graph is the total distance travelled by the particle for that part of the motion.

■ **Gradient of the speed-time graph = acceleration of the particle.**
■ **Area under the graph = distance travelled by the particle.**

Example 7
A cyclist leaves home O and rides along a straight road with a constant acceleration. After 10 seconds he has reached point A with a speed of $15\,\mathrm{m\,s^{-1}}$ and he maintains this speed for a further 20 seconds until he reaches B before retarding uniformly to rest at C. The whole journey takes 45 seconds. Sketch the speed-time graph for the journey and find:

(a) his acceleration for the first part of the journey from O to A
(b) his retardation for the final part of the journey from B to C
(c) the total distance travelled from A to C.

A sketch graph should be drawn on ordinary paper – not on graph paper – but some scales should be put on the axes.

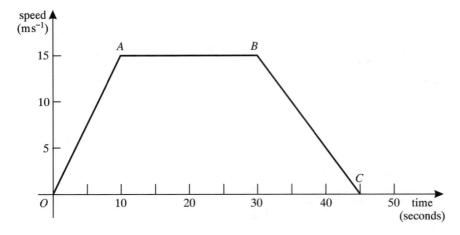

(a) The acceleration for the first part of the journey is given by the gradient of OA.

$$\text{Gradient of } OA = \frac{15}{10} = 1.5$$

The acceleration is $1.5\,\mathrm{m\,s}^{-2}$.

(b) The acceleration for the final part of the motion is the gradient of BC.

$$\text{Gradient of } BC = \frac{-15}{15} = -1$$

So the retardation is $1\,\mathrm{m\,s}^{-2}$.

(c) The total distance travelled is given by the area under the graph.

$$\text{Area of trapezium } OABC = \tfrac{1}{2}(20 + 45) \times 15$$
$$= 487.5$$

The total distance travelled is $487.5\,\mathrm{m}$.

Example 8

A car travelling along a straight road passes point A when $t = 0$ and maintains a constant speed until $t = 24$ seconds. The driver then applies the brakes and the car retards uniformly to rest at point B. Before the brakes were applied the car had travelled $\frac{4}{5}$ of the total distance AB. Sketch the speed-time graph for the journey and calculate the time taken for the car to travel from A to B.

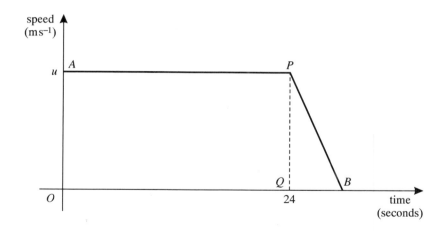

Let the initial speed be u m s^{-1}.

Then the distance in metres travelled at constant speed is:

$$\text{Area } OAPQ = (24u)$$

This is $\frac{4}{5}$ of the total distance.

As: $\text{Area } OAPQ = \frac{4}{5} \text{ area } OAPB$

And: $\text{Area } \triangle PBQ = \frac{1}{5} \text{ area } OAPB$

Then: $\triangle PBQ = \frac{1}{4} \text{area } OAPQ$

$$= \frac{1}{4} \times 24u = 6u$$

Let the car come to rest at time $(24 + T)$ seconds, then QB represents T seconds on the speed-time graph.

Then: $\text{Area } \triangle PBQ = \frac{1}{2}uT$

Hence: $\frac{1}{2}uT = 6u$

$$T = 12$$

The total time taken for the car to travel from A to B is $(24 + 12)$ s = 36 s.

Exercise 3C

1

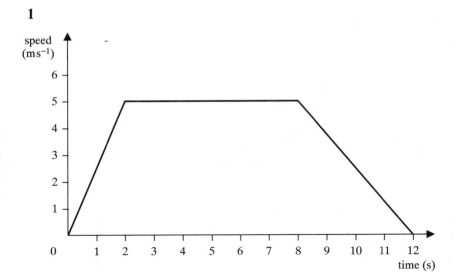

The speed-time graph above shows the speed of a particle which accelerates uniformly for 2 seconds then moves with constant speed for 6 seconds before retarding to rest. Find (a) the acceleration of the particle (b) the retardation of the particle (c) the distance it travels during the whole motion.

2 A car is moving with a speed of $20 \, \text{m s}^{-1}$ when the driver sees a red light ahead. He applies the brakes and stops in a distance of 30 m. Sketch a speed-time graph and find (a) the time taken to come to rest (b) the retardation of the car.

3 A car accelerates uniformly from rest with an acceleration of $0.9 \, \text{m s}^{-2}$ for 12 seconds. It then moves with constant speed for 16 seconds before retarding uniformly to rest. The total distance travelled is 389 m. Show this information on a speed-time graph and find (a) the greatest speed of the car (b) the total time for the journey.

4 Two stations A and B are 495 m apart. A train starts from A and accelerates uniformly for 15 s to a speed of $54 \, \text{km h}^{-1}$ which it maintains until it is 90 m from B. At this point it slows down uniformly to stop at B. Sketch a speed-time graph to show the motion of the train. Find the magnitude of the acceleration and retardation and the total time the journey takes.

5 A car passes point A on a straight road while travelling at a constant speed of $20\,\mathrm{m\,s^{-1}}$. Forty seconds later at point B the driver applies the brakes and the car comes to rest at point C. Given that the distance from B to C is $\frac{1}{6}$ of the total distance AC sketch a speed-time graph to show the motion of the car and find the total time taken to travel from A to C.

6 A train starts from rest and moves with uniform acceleration for 5 minutes. It then maintains a constant speed for 20 minutes before being brought to rest by a uniform retardation of magnitude twice that of the acceleration. Sketch a speed-time graph of the motion and find the time it takes the train to stop. If the train travels 4.5 km while accelerating show that the acceleration is $\frac{1}{10}\mathrm{m\,s^{-2}}$ and find the total distance travelled.

3.4 Velocity and acceleration when displacement is a function of time

To study this section you need to know about the differentiation and integration of x^p, where $p \neq -1$ for integration. This is dealt with in Book P1, chapters 8 and 9.

The acceleration of a moving particle is not always constant. If the acceleration is variable then calculus must be used to find the velocity of and the distance travelled by a particle. When a particle is travelling in a straight line and the displacement x measured from a fixed point on that line is given by:

$$x = f(t)$$

the velocity, which is the rate of change of the displacement with respect to time, is given by:

$$v = \frac{\mathrm{d}x}{\mathrm{d}t}$$

Acceleration is the rate of change of velocity with respect to time and is given by:

$$a = \frac{\mathrm{d}v}{\mathrm{d}t}$$

Substituting $v = \dfrac{\mathrm{d}x}{\mathrm{d}t}$ gives:

$$a = \frac{\mathrm{d}^2 x}{\mathrm{d}t^2}$$

Example 9

A particle P moves in a straight line such that its displacement x m from a fixed point O at time t seconds is given by $x = 2 + 3t - 2t^3$. Find:

(a) the distance of P from O when $t = 0$
(b) the distance of P from O when $t = 2$
(c) the velocity of P when $t = 0$
(d) the acceleration when $t = 1$.

(a) You know that: $\qquad x = 2 + 3t - 2t^3$

Substituting $t = 0$ in this expression gives $x = 2$.

When $t = 0$ the particle is 2 m from O.

(b) Substituting $t = 2$ gives:

$$x = 2 + (3 \times 2) - (2 \times 2^3) = 2 + 6 - 16 = -8$$

The **displacement** of the particle from O is -8 m when $t = 2$. This means that the distance of the particle from O is 8 m but it is now on the opposite side of O from its position when $t = 0$.

(c) Using: $\qquad x = 2 + 3t - 2t^3$

Let the velocity of P be v m s^{-1}.
Differentiating x with respect to t gives:

$$v = \frac{\mathrm{d}x}{\mathrm{d}t} = 3 - 6t^2$$

Substituting $t = 0$ gives $v = 3$.

So the velocity is 3 m s^{-1} when $t = 0$.

(d) Using: $\qquad v = 3 - 6t^2$

If the acceleration is a m s^{-2} then:

$$a = \frac{\mathrm{d}v}{\mathrm{d}t} = -12t$$

Substituting $t = 1$ gives $a = -12$.
The acceleration of the particle is -12 m s^{-2} when $t = 1$. It has a retardation of 12 m s^{-2}.

Determining the velocity and displacement when the acceleration or velocity is given

If the acceleration is given as a function of time, then you need to reverse the process to find the velocity and displacement. In other words, you need to use integration.

Suppose at time t seconds the acceleration is a m s^{-2}, the velocity v m s^{-1} and the displacement x m.

Since:
$$a = \frac{dv}{dt} = f(t)$$

Integrating with respect to time gives:
$$v = \int f(t)\ dt + c$$

where c is the constant of integration.

Similarly if:
$$v = g(t)$$

then:
$$x = \int g(t)\ dt + k$$

where k is the constant of integration.

Example 10

A particle P moves in a straight line so that at time t seconds its acceleration is $3t$ m s^{-2}. Given that P passes through the point O at time $t = 0$ with velocity 2 m s^{-1}, find

(a) the velocity of P in terms of t

(b) the distance of P from O when $t = 3$.

(a) Since:
$$a = \frac{dv}{dt} = 3t$$

Integrating with respect to time gives:
$$v = \tfrac{3}{2}t^2 + c$$

Since $t = 0$ when $v = 2$, substituting these values into the equation gives $c = 2$.

So the velocity of P is given by:
$$v = \tfrac{3}{2}t^2 + 2$$

(b) You now know that:
$$v = \frac{dx}{dt} = \tfrac{3}{2}t^3 + 2$$

Integrating with respect to time gives:
$$x = \tfrac{3}{3} \times \frac{t^3}{2} + 2t + k$$
$$= \frac{t^3}{2} + 2t + k$$

You know that when $t = 0$ $x = 0$.

Substituting these values gives $k = 0$.

So:
$$x = \frac{t^3}{2} + 2t$$

You need to find x when $t = 3$.

Substituting $t = 3$ gives:

$$x = \frac{3^3}{2} + 2 \times 3$$
$$= 13\tfrac{1}{2} + 6$$
$$= 19\tfrac{1}{2}$$

The particle is $19\tfrac{1}{2}$ m from O when $t = 3$.

Example 11

A particle moves in a straight line so that at time t seconds its velocity is given by $(2t^2 - 5t + 3)\,\mathrm{m\,s^{-1}}$. Initially its displacement from the fixed point O on the line is $5\,\mathrm{m}$. Find:
(a) the times when the particle is at rest
(b) the displacement of the particle from O when $t = 3$.

(a) You know that $\qquad v = 2t^2 - 5t + 3$

The particle is at rest when $v = 0$.

Hence it is at rest when $0 = 2t^2 - 5t + 3$.

Factorising gives:

$$0 = (2t - 3)(t - 1)$$
So either: $\qquad t = \tfrac{3}{2}$
Or: $\qquad t = 1$

The particle is at rest when $t = 1$ s and $t = 1\tfrac{1}{2}$ s.

(b) The displacement of the particle can be found by using:

$$v = \frac{\mathrm{d}x}{\mathrm{d}t} = 2t^2 - 5t + 3$$

Integrating with respect to time t gives:

$$x = \tfrac{2}{3}t^3 - \tfrac{5}{2}t^2 + 3t + c$$

Initially that is when $t = 0$, the particle's displacement from the fixed point O on the line is $5\,\mathrm{m}$.

Substituting $t = 0$, $x = 5$ gives $c = 5$.

So: $\qquad x = \frac{2}{3}t^3 - \frac{5}{2}t^2 + 3t + 5$

Substituting $t = 3$ gives: $\quad x = \left(\frac{2}{3} \times 3^3\right) - \left(\frac{5}{2} \times 3^2\right) + (3 \times 3 + 5)$

$$= 18 - 22\frac{1}{2} + 9 + 5 = 9\frac{1}{2}$$

The particle is $9\frac{1}{2}$ m from O when $t = 3$.

Motion in a plane

Studying the motion of a particle in two dimensions in the (x, y) plane involves using vectors. If unit vectors **i** and **j** are taken in the directions of the x-and y-axes respectively then the position vector **r** m relative to a fixed origin O in the plane of a moving particle P may be given by:

$$\mathbf{r} = \mathrm{f}(t)\mathbf{i} + \mathrm{g}(t)\mathbf{j}$$

where $\mathrm{f}(t)$ and $\mathrm{g}(t)$ are both functions of time t.

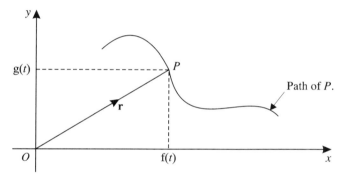

The velocity and acceleration of P can be found by differentiating each component separately, as shown in chapter 2. With the usual notation of velocity being $\mathbf{v}\,\mathrm{m\,s}^{-1}$ and acceleration $\mathbf{a}\,\mathrm{m\,s}^{-2}$:

$$\mathbf{v} = \frac{\mathrm{d}\mathbf{r}}{\mathrm{d}t} \quad \text{and} \quad \mathbf{a} = \frac{\mathrm{d}\mathbf{v}}{\mathrm{d}t} = \frac{\mathrm{d}^2\mathbf{r}}{\mathrm{d}t^2}$$

As: $\qquad\qquad \mathbf{r} = \mathrm{f}(t)\,\mathbf{i} + \mathrm{g}(t)\mathbf{i}$

it follows that: $\qquad \mathbf{v} = \dfrac{\mathrm{d}\mathrm{f}(t)}{\mathrm{d}t}\mathbf{i} + \dfrac{\mathrm{d}\mathrm{g}(t)}{\mathrm{d}t}\mathbf{j}$

and: $\qquad\qquad \mathbf{a} = \dfrac{\mathrm{d}^2\mathrm{f}(t)}{\mathrm{d}t^2}\mathbf{i} + \dfrac{\mathrm{d}^2\mathrm{g}(t)}{\mathrm{d}t^2}\mathbf{j}$

When integrating include the constant of integration in vector form.

Example 12

The position vector **r** metres of a particle P relative to a fixed point O is given by $\mathbf{r} = (5t^2 - 2)\mathbf{i} + (t - 3)\mathbf{j}$. Find the velocity and acceleration in vector form.

As:
$$\mathbf{r} = (5t^2 - 2t)\mathbf{i} + (t - 3)\mathbf{j}$$

it follows that:
$$\mathbf{v} = \frac{d\mathbf{r}}{dt} = (10t - 2)\mathbf{i} + \mathbf{j}$$

and:
$$\mathbf{a} = \frac{d\mathbf{v}}{dt} = 10\mathbf{i}$$

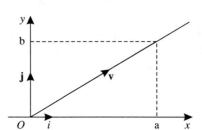

The velocity vector is $[(10t - 2)\mathbf{i} + \mathbf{j}]\,\mathrm{m\,s^{-1}}$ and the acceleration vector is $10\mathbf{i}\,\mathrm{m\,s^{-2}}$.

The speed of a particle is the **magnitude** of its velocity. (Speed is a scalar quantity not a vector.)

$$\text{speed} = |\text{velocity}|$$

If $\mathbf{v} = a\mathbf{i} + b\mathbf{j}$ where a and b are constants then by Pythagoras' Theorem $|\mathbf{v}| = \sqrt{(a^2 + b^2)}$.

■
$$\text{speed} = \sqrt{(a^2 + b^2)}\,\mathrm{m\,s^{-1}}$$

Example 13

A particle P moves so that its acceleration $\mathbf{a}\,\mathrm{m\,s^{-2}}$ at time t seconds is given by $\mathbf{a} = 2t\mathbf{i} + 4\mathbf{j}$. When $t = 0$ the velocity of P is $(3\mathbf{i} - 7\mathbf{j})\,\mathrm{m\,s^{-1}}$. Find the speed of P when $t = 3$.

$$\mathbf{a} = \frac{d\mathbf{v}}{dt} = 2t\mathbf{i} + 4\mathbf{j}$$

By integration $\mathbf{v} = t^2\mathbf{i} + 4t\mathbf{j} + A\mathbf{i} + B\mathbf{j}$ where A and B are scalar constants.

When $t = 0$:
$$\mathbf{v} = 3\mathbf{i} - 7\mathbf{j}$$

So:
$$3\mathbf{i} - 7\mathbf{j} = A\mathbf{i} + B\mathbf{j}$$

Hence:
$$\mathbf{v} = t^2\mathbf{i} + 4t\mathbf{j} + 3\mathbf{i} - 7\mathbf{j}$$
$$\mathbf{v} = (t^2 + 3)\mathbf{i} + (4t - 7)\mathbf{j}$$

Substituting $t = 3$ gives:
$$\mathbf{v} = (3^2 + 3)\mathbf{i} + (4 \times 3 - 7)\mathbf{j}$$
$$= 12\mathbf{i} + 5\mathbf{j}$$

$$|\mathbf{v}| = \sqrt{(12^2 + 5^2)} = 13$$

The particle's speed when $t = 3$ is $13\,\mathrm{m\,s^{-1}}$.

Exercise 3D

1 A particle moves in a straight line such that at time t its displacement x from a fixed point O on that line is given by $x = 3t^3 - 2t^2$. Find (a) its velocity when $t = 2$ and (b) its acceleration when $t = 1$.

2 A particle moves in a straight line such that at time t its displacement x from a fixed point O on that line is given by $x = 3t^3 - 2t^2$. Find (a) the times when it is at rest and (b) its distances from O when it is at rest.

3 A particle moves in a straight line with velocity v given by $v = 3t - 4t^3 + 1$ at time t s after passing through a fixed point O in that line. Find (a) its velocity when $t = 1$ (b) its acceleration when $t = 2$ (c) its maximum velocity (d) its distance from O when $t = 2$.

4 A particle P moves in straight line such that at time t s its acceleration is given by $a = 2t - 2$. It passes through a point O on the line when $t = 0$ with a velocity of $1\,\mathrm{m\,s^{-1}}$. Find (a) the velocity of P when $t = 2$ (b) the displacement of P from O when $t = 2$ (c) the time when P is at rest (d) the distance of P from O when it is at rest.

5 A particle P is moving along the x-axis with velocity $v = 5t^2 - 3t$. When $t = 0$ $x = 4$. Find (a) the times when P is at rest (b) the position of P each time it is at rest (c) the acceleration of P when $t = 2$.

6 A particle P is moving along the x-axis with velocity $v = 4t - 2t^2$. When $t = 0$ P is at $x = 3$. Find (a) the position of P when $t = 2$ (b) the maximum velocity attained by P (c) the distance OP when the velocity is maximum.

7 A particle P is moving along the x-axis. When $t = 0$ the velocity of P is $4\frac{1}{2}\,\mathrm{m\,s^{-1}}$. At time t s the acceleration of P is given by $(3t - 6)\,\mathrm{m\,s^{-2}}$. Find the time when the particle returns to its starting point.

8 A particle moves along the x-axis with velocity at time t given by $v = 5t^2 + 2t$. Find (a) the distance it moves in the 2nd second (b) the distance it moves in the 4th second.

9 The position vector $\mathbf{r}$ m of a particle P relative to a fixed point O is given by $\mathbf{r} = (3t^2 - 4t)\mathbf{i} + (2t^2 - 1)\mathbf{j}$. Find the velocity vector of P and show that the acceleration of P is constant.

10 A particle P is moving in the (x, y) plane with velocity $\mathbf{v}\,\mathrm{m\,s^{-1}}$
at time t seconds given by $\mathbf{v} = 2t^2\mathbf{i} + 2t^{\frac{3}{2}}\mathbf{j}$.
   (a) Find the acceleration of P when $t = 4$. Given that P is at
   O when $t = 0$
   (b) Find the position vector of P relative to O when $t = 1$
   (c) Find the distance of P from O when $t = 1$.

11 A particle P starts from rest at a fixed origin O when $t = 0$.
The acceleration $\mathbf{a}\,\mathrm{m\,s^{-2}}$ of P at time t seconds is given by
$\mathbf{a} = 6t\mathbf{i} - 4\mathbf{j}$. Find:
   (a) the velocity vector of P when $t = 2$
   (b) the speed of P when $t = 2$
   (c) the position vectors of P relative to O when $t = 1$ and $t = 2$
   (d) the distance between the positions of P when $t = 1$ and
   $t = 2$.

12 A particle moves with acceleration $\mathbf{a}\,\mathrm{m\,s^{-2}}$ at time t seconds
given by $\mathbf{a} = 7t\mathbf{i} + 4t^2\mathbf{j}$. Given that $\mathbf{v} = 3\mathbf{i} - 4\mathbf{j}$ when $t = 0$
find the speed of the particle when $t = 2$.

13 When $t = 0$ a particle P is at rest at point A whose position
vector relative to a fixed point O is $\mathbf{r} = (3\mathbf{i} + 4\mathbf{j})\,\mathrm{m}$. P moves
with acceleration $\mathbf{a}\,\mathrm{m\,s^{-2}}$ at time t seconds given by $\mathbf{a} = 5t\mathbf{i}$.
Find the distance P moves between $t = 1$ and $t = 2$.

14 When $t = 0$ a particle P is at the point with position vector
$(3\mathbf{i} + 4\,\mathbf{j})\,\mathrm{m}$ relative to a fixed point O. The velocity $\mathbf{v}\,\mathrm{m\,s^{-1}}$
at time t seconds of P is given by $\mathbf{v} = 6t^2\mathbf{i} + (2t + 3)\mathbf{j}$.
   (a) Find the position vector of P relative to O at time t
   seconds. When $t = 0$ a second particle Q is at the point with
   position vector $(2\mathbf{i} + 10\mathbf{j})\,\mathrm{m}$ relative to O and the velocity
   $\mathbf{v}\,\mathrm{m\,s^{-1}}$ at time t seconds for Q is given by $\mathbf{v} = 3\mathbf{i} - 2\mathbf{j}$.
   (b) Find the position vector of Q relative to O at time t
   seconds.
   (c) Show that P and Q will collide when $t = 1$ and find the
   position vector of their point of collision.

15 A particle P moves so that at time t seconds its position
vector $\mathbf{r}\,\mathrm{m}$ relative to a fixed origin O is given by:

$$\mathbf{r} = (2t^2 - pt)\mathbf{i} + (t^3 - 3t)\mathbf{j}$$

where p is a constant.

(a) Find an expression for the velocity, $\mathbf{v}\,\mathrm{m\,s}^{-1}$, at time t seconds.

(b) Given that the particle comes to instantaneous rest find the time at which it is at rest and hence the value of p.

16 At time t seconds two particles A and B have position vectors $\mathbf{r}_a$ m and $\mathbf{r}_b$ m respectively with respect to a fixed origin O where:

$$\mathbf{r}_a = (t^2 + 12t)\mathbf{i} + 4t\mathbf{j}$$
$$\mathbf{r}_b = t^2\mathbf{i} + (2t - 6)\mathbf{j}$$

Find the velocity vectors $\mathbf{v}_a\,\mathrm{m\,s}^{-1}$ and $\mathbf{v}_b\,\mathrm{m\,s}^{-1}$ of A and B at time t seconds.

Hence determine the value of t for which the speed of A is twice the speed of B.

17 At time t seconds two particles P and Q have position vectors $\mathbf{r}_p$ m and $\mathbf{r}_q$ m respectively relative to a fixed origin O where:

$$\mathbf{r}_p = \tfrac{1}{2}t^2\mathbf{i} + 2\mathbf{j} \quad \text{and} \quad \mathbf{r}_q = 2t(3\mathbf{i} - 4\mathbf{j})$$

(a) Calculate the distance between P and Q when $t = 2$

(b) Show that the velocity of Q is constant and calculate its magnitude and direction

(c) Calculate the acceleration of P.

3.5 Projectiles

Ignoring air resistance, a particle thrown into the air moves freely under gravity. Such a particle is generally called a **projectile**. It moves through the air along a curved path.

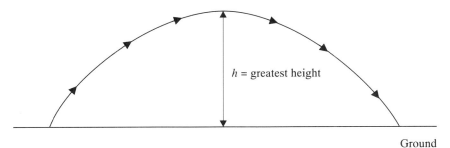

h = greatest height

Ground

A projectile has an acceleration of $g = 9.8\,\mathrm{m\,s^{-2}}$ vertically downwards. Problems involving projectiles can be solved by considering the vertical and horizontal motions separately. As the acceleration is entirely vertical the projectile has no acceleration horizontally and so its horizontal velocity is unchanged throughout the motion. Vertically, the motion can be investigated by applying the uniform acceleration equations as in section 3.2.

Horizontal projection

A stone which is thrown horizontally from the top of a cliff will fall into the sea below, but will travel horizontally as well as vertically while falling.

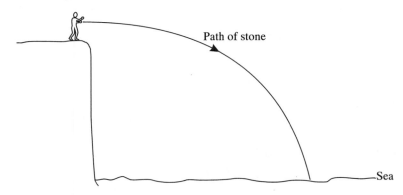

Path of stone

Sea

The stone begins its journey at the highest point of the projectile path shown in the previous diagram. At this highest point the stone's vertical velocity is zero but it has a vertical acceleration, g.

Example 14

A particle is projected horizontally with speed $15\,\mathrm{m\,s^{-1}}$ from a point 24 m above a horizontal surface. Calculate to three significant figures the time taken for the particle to reach the surface and the horizontal distance travelled in that time.

Since the vertical distance to the plane is known, use the vertical motion to find the time taken to reach the surface. As in section 3.2 choose a direction to be positive.

The vertical motion is downwards, so take this direction to be positive.

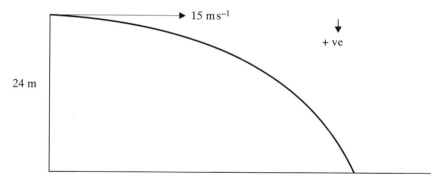

As the particle starts with a horizontal velocity its initial vertical speed is zero.

Known quantities are:
$$a = 9.8 \, \text{m s}^{-2}$$
$$\downarrow \quad s = 24 \, \text{m}$$
$$+\text{ve} \quad u = 0 \, \text{m s}^{-1}$$

Using:
$$s = ut + \tfrac{1}{2}at^2$$

Gives:
$$24 = 0 \times t + \tfrac{1}{2} \times 9.8 \times t^2$$
$$t^2 = \frac{48}{9.8}$$
$$t = 2.213$$

The time taken to reach the surface is 2.21 s.

Horizontal motion must be considered to find the horizontal distance travelled. As there is no horizontal acceleration the horizontal velocity is unchanged.

Hence:
$$\text{distance travelled} = (\text{horizontal speed}) \times \text{time}$$
$$= (15 \times 2.213)$$
$$= 33.19$$

The horizontal distance travelled is 33.2 m.

Example 15

A particle is projected horizontally with a speed of $19.6\,\mathrm{m\,s^{-1}}$. Find the distance of the particle from its starting point after 1.5 seconds.

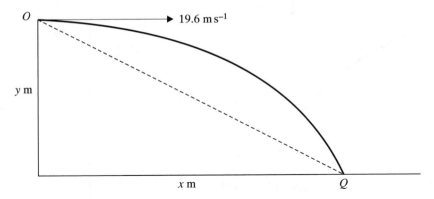

The distance between the particle's position Q when t $= 1.5$ s and its starting point O is shown by the dotted line in the above diagram. You need to find the horizontal and vertical distances, x m and y m.

Consider the horizontal motion.

As there is no horizontal acceleration:

$$x = \text{(horizontal speed)} \times \text{time}$$
$$= (19.6 \times 1.5) = 29.4$$

The distance x is 29.4 m.

Now consider the vertical motion.

Known quantities are:
$$u = 0 \text{ ms}^{-1}$$
$$\downarrow \qquad a = 9.8\,\mathrm{m\,s^{-2}}$$
$$+\text{ve} \qquad t = 1.5\,\text{s}$$

Using: $\qquad s = ut + \tfrac{1}{2}at^2$

Gives: $\qquad y = 0 \times 1.5 + \tfrac{1}{2} \times 9.8 \times 1.5^2$
$$y = 11.025$$

The distance y is 11.025 m.

By Pythagoras' Theorem:
$$OQ = \sqrt{(x^2 + y^2)} = \sqrt{(29.4^2 + 11.025^2)}$$
$$= 31.39$$

The particle is 31.4 m from its point of projection after 1.5 seconds.

Example 16

A stone is thrown horizontally with a speed of $20\,\mathrm{m\,s^{-1}}$. By modelling the stone as a particle find the magnitude and direction of the stone's velocity 2 seconds later.

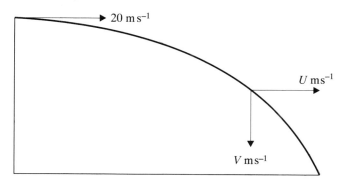

To find the magnitude and direction of the stone's velocity after 2 seconds you need the resultant of the horizontal and vertical components of the velocity.

Let the components be $U\,\mathrm{m\,s^{-1}}$ and $V\,\mathrm{m\,s^{-1}}$ as shown in the diagram.

For horizontal motion there is no acceleration so $U = 20$.

For vertical motion the known quantities are:

$$\downarrow \qquad t = 2\,\mathrm{s}$$
$$+\mathrm{ve} \qquad u = 0\,\mathrm{m\,s^{-1}}$$
$$a = 9.8\,\mathrm{m\,s^{-2}}$$
$$v = V\,\mathrm{m\,s^{-1}}$$

Using: $v = u + at$

Gives: $V = 0 + 9.8 \times 2 = 19.6$

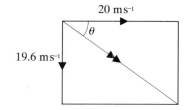

The two components of the stone's velocity are perpendicular. Their resultant is the diagonal of the rectangle shown opposite.

The magnitude of the resultant $= \sqrt{(20^2 + 19.6^2)} = 28.0$.

As: $\tan \theta = \dfrac{19.6}{20}$

Hence: $\theta = 44.4°$

The stone's velocity has magnitude $28\,\mathrm{m\,s^{-1}}$ and its direction is $44.4°$ below the horizontal.

Projectiles using vector notation

Information about a projectile can be conveyed in vector form with **i** and **j** being the unit vectors in the horizontal and upward vertical directions respectively. Separate the horizontal and vertical information and solve as before.

Example 17

A particle is projected at $t = 0$ with a velocity $5\mathbf{i}\,\mathrm{m\,s}^{-1}$ from a point with position vector $(6\mathbf{i} + 50\mathbf{j})\,\mathrm{m}$ relative to a fixed origin O where **i** and **j** are unit vectors in the horizontal and upward vertical directions respectively. Find the position vector of the particle 3 seconds later.

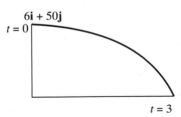

There is no horizontal acceleration.

So: $\qquad$ Distance $=$ speed $\times$ time

$$s = (5 \times 3) = 15$$

The horizontal distance travelled in vector form is $15\mathbf{i}$.

For vertical motion known quantities are:

$$\downarrow \quad a = 9.8\,\mathrm{m\,s}^{-2}$$
$$u = 0 \ \mathrm{ms}^{-1}$$

Using: $\qquad s = ut + \frac{1}{2}at^2$

Gives: $\qquad s = \frac{1}{2} \times 9.8 \times 3^2 = 44.1$

The vertical distance is downwards. In vector form it is therefore $-44.1\mathbf{j}$.

The total displacement is $15\mathbf{i} - 44.1\mathbf{j}$.

Using the vector diagram opposite gives:

$$\text{final position vector} = (6\mathbf{i} + 50\mathbf{j}) + (15\mathbf{i} - 44.1\mathbf{j})$$
$$= 21\mathbf{i} + 5.9\mathbf{j}$$

The position vector after 3 seconds is $(21\mathbf{i} + 5.9\mathbf{j})\,\mathrm{m}$.

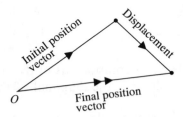

Exercise 3E

Whenever a numerical value of g is required take $g = 9.8\,\mathrm{m\,s}^{-2}$.

1 A particle is projected horizontally at $15\,\mathrm{m\,s}^{-1}$ from a point 60 m above a horizontal surface. Find how long it takes

before the particle strikes the surface and the horizontal distance travelled in that time.

2 A particle is projected horizontally at $20\,\mathrm{m\,s^{-1}}$. It strikes a horizontal surface after travelling $50\,\mathrm{m}$ horizontally. Find the height of the point of projection above the surface.

3 A stone is thrown horizontally from a window $3.6\,\mathrm{m}$ above horizontal ground. It hits the ground after travelling $10\,\mathrm{m}$ horizontally. Find the speed of projection.

4 A stone is thrown horizontally at $25\,\mathrm{m\,s^{-1}}$ from the edge of a vertical cliff which is $50\,\mathrm{m}$ high. By modelling the stone as a particle calculate its distance from the foot of the cliff when it enters the water.

5 A bird flying horizontally at $30\,\mathrm{m\,s^{-1}}$ at a height of $35\,\mathrm{m}$ drops a stone. Model the stone as a particle and hence determine how long it takes the stone to reach the ground. What is the horizontal distance travelled by the stone during this time?

6 A stone is thrown horizontally and hits the ground $7\,\mathrm{m}$ from the thrower $0.5\,\mathrm{s}$ later. Find the speed and height of projection.

7 At time $t = 0$ a particle is projected with a velocity $4\mathbf{i}\,\mathrm{m\,s^{-1}}$ from a point with position vector $24\mathbf{j}\,\mathrm{m}$ where $\mathbf{i}$ and $\mathbf{j}$ are unit vectors in the horizontal and upward vertical directions respectively. Find the position vector of the particle after $1.5\,\mathrm{s}$.

8 At time $t = 0$ a particle is projected horizontally from a point with position vector $(x\mathbf{i} + y\mathbf{j})\,\mathrm{m}$. Two seconds later it passes through the point with position vector $(8\mathbf{i} + 2\mathbf{j})\,\mathrm{m}$. If the speed of projection is $4\,\mathrm{m\,s^{-1}}$ find the values of x and y.

9 A child throws a ball horizontally at a wall $4\,\mathrm{m}$ away. The ball strikes the wall $0.4\,\mathrm{m}$ below the level of projection. Calculate the speed with which it was projected.

10 A particle slides along a horizontal table and over the edge. Given that the table is $1\,\mathrm{m}$ high and the speed of the particle on the table is $14\,\mathrm{m\,s^{-1}}$ calculate (a) the vertical velocity with which the particle hits the floor (assumed horizontal) (b) the

speed and direction of motion of the particle when it hits the floor.

11 A boy standing on a garage roof throws a ball horizontally with a speed of $15 \, \text{m s}^{-1}$. The ball just clears a wall 1.5 m high 10 m away. Model the ball as a particle and hence calculate the height of projection.

12 A particle is projected horizontally at $25 \, \text{m s}^{-1}$. Find the horizontal and vertical components of the particle's velocity 1.5s later. Find the speed of the particle and its direction of motion at that time.

13 At time $t = 0$ a particle is projected with a velocity $4\mathbf{i} \, \text{m s}^{-1}$ from a point with position vector $(6\mathbf{i} + 2\mathbf{j}) \, \text{m}$. Find its position vector after 2 seconds.

14 At time $t = 0$ a particle is projected horizontally from a point A with position vector $(5\mathbf{i} + 30\mathbf{j}) \, \text{m}$. It passes through point B with position vector $(17\mathbf{i} + 10.4\mathbf{j}) \, \text{m}$. Find its speed of projection and the time taken to travel from A to B.

15 An aeroplane is flying horizontally at $400 \, \text{m s}^{-1}$. A package is released and travels a distance of 2000 m horizontally before hitting the ground. Model the package as a particle and hence find the height of the aeroplane above the ground.

16 A batsman hits a ball horizontally with a speed of $21 \, \text{m s}^{-1}$ at 1 m above the ground. Find the distance travelled horizontally by the ball before it reaches the ground.

17 A child throws a ball horizontally from a window 5 m above horizontal ground. The ball just clears a vertical wall 2.5 m high and 12 m from the house. By modelling the ball as a particle calculate the speed of projection. The ball hits the ground at Q. Calculate the distance of Q from the wall.

18 A tennis ball is served horizontally at a speed of $24 \, \text{m s}^{-1}$ from a height of 2.7 m. The net is 1 m high and 12 m horizontally from the server. Model the ball as a particle and hence determine whether the ball clears the net and if so by what distance.

Projection at an angle to the horizontal

It is more usual for particles to be projected at an angle to the horizontal rather than horizontally.

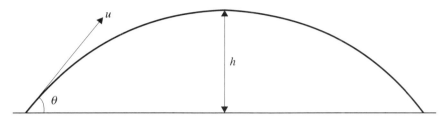

It is easier to investigate the motion of such a particle if you deal with its horizontal and vertical movement separately.

If a particle is projected with initial speed u at an angle θ to the horizontal the horizontal component of the initial velocity can be expressed as $u \cos \theta$. This remains constant throughout the motion. The vertical component of the initial velocity is $u \sin \theta$ which does not remain constant because of the vertical acceleration due to gravity.

- **For initial velocity:**   **horizontal component $= u \cos \theta$**
  and: **vertical component $= u \sin \theta$.**

Example 18

A particle is projected from a point O with a speed of $30 \, \mathrm{m \, s^{-1}}$ at an angle of elevation of $\arcsin \frac{3}{5}$ (in other words the angle of elevation is θ where $\sin \theta = \frac{3}{5}$).

(a) Find the greatest height above O reached by the particle
(b) The particle strikes the horizontal through O at Q. Find the distance OQ.

(a) The vertical speed when the particle reaches its greatest height is zero.

Its initial speed is:

$$(30 \sin \theta) \, \mathrm{m \, s^{-1}} = (30 \times \tfrac{3}{5}) \, \mathrm{m \, s^{-1}} = 18 \, \mathrm{m \, s^{-1}}$$

Known quantities for vertical motion are:

$$+\text{ve} \quad u = 18 \, \mathrm{m \, s^{-1}}$$
$$\uparrow \quad v = 0 \, \mathrm{m \, s^{-1}}$$
$$a = -9.8 \, \mathrm{m \, s^{-2}}$$

Let the greatest height be h m.

Using: $\qquad v^2 = u^2 + 2as$

Gives: $\qquad 0 = 18^2 + 2 \times (-9.8)h$

$$h = \frac{18^2}{2 \times 9.8} = 16.53$$

The greatest height is 16.5 m.

(b) To calculate the horizontal distance you need both the velocity and the time taken. When the particle reaches Q, its vertical displacement from its starting point is zero. Use this to find the length of time the particle is in the air – often called **the time of flight**.

To find this time of flight T, consider the vertical motion.

Known quantities for vertical motion are:

$$+\text{ve} \qquad u = 18 \, \text{m s}^{-1}$$
$$\uparrow \qquad s = 0 \, \text{m}$$
$$a = -9.8 \, \text{m s}^{-2}$$

Using: $\qquad s = ut + \frac{1}{2}at^2$

Gives: $\qquad 0 = 18t - \frac{1}{2} \times 9.8t^2$

Factorising gives: $\qquad 0 = t(18 - 4.9t)$

So either: $t = 0$ $\qquad$ (this is the moment of projection)

or: $t = \dfrac{18}{4.9} = 3.673$ $\qquad$ (this is the time when the particle reaches Q.)

It takes 3.673 s to travel from O to Q.
Since there is no horizontal acceleration:

$$\text{horizontal distance} = \text{speed} \times \text{time}$$

The horizontal velocity is $30 \cos \theta \, \text{m s}^{-1}$.

As $\sin \theta = \frac{3}{5}$, a (3, 4, 5) triangle shows that $\cos \theta = \frac{4}{5}$.

So: $\qquad OQ = (30 \cos \theta \times 3.673)$
$$= (30 \times \tfrac{4}{5} \times 3.673)$$
$$= 88.16$$

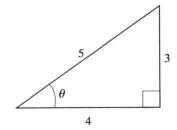

The distance travelled horizontally is 88.2 m. This distance is horizontal often called the **range of the projectile**.

Example 19

A girl hits a ball at an angle of arctan $\frac{3}{4}$ to the horizontal (that is at an angle θ where $\tan\theta = \frac{3}{4}$) from a point O which is 0.5 m above level ground. The initial speed of the ball is $15\,\text{m}\,\text{s}^{-1}$. The ball just clears a fence which is a horizontal distance of 18 m from the girl, as shown in the diagram. By modelling the ball as a projectile find the time taken for the ball to reach the fence and the height of the fence.

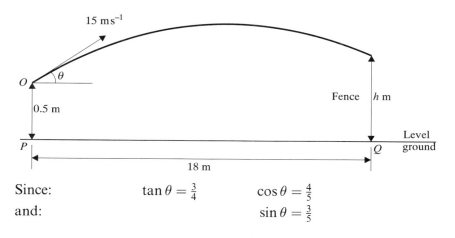

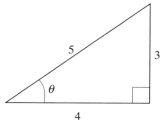

Since: $\tan\theta = \frac{3}{4}$ $\qquad$ $\cos\theta = \frac{4}{5}$

and: $\qquad\qquad\qquad\qquad$ $\sin\theta = \frac{3}{5}$

In this example, the horizontal speed and distance PQ are known, so the time to reach the top of the fence can be found.

The horizontal speed is $(15\cos\theta) = (15 \times \frac{4}{5}) = 12\,\text{m}\,\text{s}^{-1}$.

Since horizontal distance $= 18$ m:

$$\text{time} = \frac{\text{distance}}{\text{speed}} = \frac{18}{12} = 1.5$$

The ball takes 1.5 s to reach the fence.

Let the fence be h metres high. At the moment when the ball just clears the fence it is $(h - 0.5)$ m higher than its starting point.

Known quantities for vertical motion are:

$\qquad$ + ve $\qquad$ $s = (h - 0.5)\,\text{m}$

$\qquad$ $\uparrow$ $\qquad$ $u = 15\sin\theta\,\text{m}\,\text{s}^{-1} = 15 \times \frac{3}{5}\,\text{m}\,\text{s}^{-1} = 9\,\text{m}\,\text{s}^{-1}$

$\qquad\qquad\qquad$ $t = 1.5\,\text{s}$

$\qquad\qquad\qquad$ $a = -9.8\,\text{m}\,\text{s}^{-2}$

Using: $$s = ut + \tfrac{1}{2}at^2$$

Gives: $$h - 0.5 = 9 \times 1.5 + \tfrac{1}{2}(-9.8) \times 1.5^2$$
$$h = 0.5 + 9 \times 1.5 - 4.9 \times 1.5^2$$
$$h = 2.95$$

The fence is 2.95 m high.

Example 20

A ball is projected from O at the top of a cliff with speed $V\,\mathrm{m\,s}^{-1}$ at an angle of $30°$ above the horizontal. The greatest height reached above the horizontal plane through O is 4.9 m. Show that $V = 19.6$. Given that O is 60 m above the surface of the sea, find the time taken for the ball to travel from O to the sea's surface.

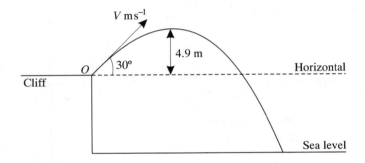

Consider the vertical motion to the highest point.

At this point, $s = 4.9$ m.

The initial vertical speed is $V \sin 30\,\mathrm{m\,s}^{-1} = \tfrac{1}{2}V\,\mathrm{m\,s}^{-1}$.

At the highest point $v = 0$.

Using: $$v^2 = u^2 + 2as \quad \text{with} \quad a = -9.8\,\mathrm{m\,s}^{-2}$$

Gives: $$0 = (\tfrac{1}{2}V)^2 + 2 \times (-9.8) \times 4.9$$
$$(\tfrac{1}{2}V)^2 = 2 \times 9.8 \times 4.9$$
$$\tfrac{1}{2}V = 9.8$$
$$V = 19.6$$

Consider now the vertical motion to sea level.

At sea level, the ball is 60 m below 0, so s $= -60$ m.

The initial vertical speed is: $\tfrac{1}{2}V\,\mathrm{m\,s}^{-1} = 9.8\,\mathrm{m\,s}^{-1}$

Using: $s = ut + \frac{1}{2}at^2$ with a $= -9.8\,\mathrm{m\,s}^{-2}$

gives: $-60 = 9.8t - \frac{1}{2} \times 9.8t^2$

$4.9t^2 - 9.8t - 60 = 0$

This quadratic does not factorise so use the quadratic formula as shown in Book P1, chapter 2 to obtain

$$t = \frac{9.8 \pm \sqrt{(9.8^2 + 4 \times 4.9 \times 60)}}{2 \times 4.9}$$

$$t = \frac{9.8 \pm 35.66}{9.8}$$

The solution $t = \dfrac{9.8 - 35.66}{9.8}$ gives t negative which is not possible.

The other solution is $t = \dfrac{9.8 + 35.66}{9.8} = 4.639$

The time taken to reach the sea is 4.64 s.

Exercise 3F

Whenever a numerical value of g is required take g $= 9.8\,\mathrm{m\,s}^{-2}$.

1 A particle is projected with a speed of 49 m s^{-1} at an angle of 45° above the horizontal. Find (a) the time taken by the particle to reach its maximum height (b) the maximum height reached (c) the time of flight (d) the horizontal range of the particle.

2 A particle is projected with a speed of 56 m s^{-1} at an angle of 30° above the horizontal. Find (a) the time taken by the particle to reach its maximum height (b) the maximum height reached (c) the time of flight (d) the horizontal range of the particle.

3 A particle is projected with a speed of 28 m s^{-1} at an angle α above the horizontal. If the greatest height reached above the point of projection is 14 m find the value of α.

4 A particle is projected with a speed of 21 m s^{-1} at an angle of elevation of 60°. Find its speed and direction of motion after (a) 1 second (b) 2 seconds (c) 3 seconds.

5 A ball is thrown from O with a speed of 28 m s^{-1} at an angle of elevation of 60°. It hits a wall which is 5 m horizontally

from O. By modelling the ball as a particle calculate the height above O at which the ball hits the wall.

6 A ball is thrown at $14\,\text{m s}^{-1}$ at an angle of elevation of $60°$. Find its horizontal and vertical distances from the point of projection after 1 second. Find the direction of motion of the ball at this time.

7 A particle is projected from the origin at a velocity of $(5\mathbf{i} + 16\mathbf{j})\,\text{m s}^{-1}$ where $\mathbf{i}$ and $\mathbf{j}$ are unit vectors in the horizontal and upward vertical directions respectively. Find the position vector of the particle after 2 seconds. Find its distance from the origin after 3 seconds.

8 A particle is projected from the origin at a velocity of $(4\mathbf{i} + 21\mathbf{j})\,\text{m s}^{-1}$ where $\mathbf{i}$ and $\mathbf{j}$ are unit vectors in the horizontal and upward vertical directions respectively. Find its position and velocity vectors after 2 seconds.

9 A golfer hits a golf ball resting on a tee with a velocity of $49\,\text{m s}^{-1}$ at an angle θ above the horizontal where $\sin \theta = \frac{3}{5}$. By modelling the ball as a particle and the ground as horizontal find the distance the ball travels before it hits the ground for the first time. What assumption have you made about the point of projection when using this model?

10 A stone is thrown at an angle of elevation of $30°$. If 1 s later it hits the ground 1 m below its point of projection find the speed of projection. Find its greatest height above the point of projection.

11 A ball is thrown from O with a speed of $30\,\text{m s}^{-1}$ at an angle of $30°$ above the horizontal. Given that it just clears the top of a wall at a horizontal distance of 20 m from O, find the height of the top of the wall above O.

12 A gun fires a shell with an initial velocity of $770\,\text{m s}^{-1}$ at an angle of $18°$ above the horizontal. By modelling the shell as a particle find the range of the shell.

13 A stone is thrown from the top of a cliff which is 80 m above sea level. Initially the stone is moving at $20\,\text{m s}^{-1}$ at an angle of elevation of $\arcsin \frac{3}{5}$. Find the time taken by the stone to reach the surface of the sea and the horizontal distance travelled in this time.

14 A ball is hit by a racquet which is almost touching the ground. Two seconds later the ball just clears a vertical wall 4 m away horizontally and 4 m high. By modelling the ball as a particle projected at ground level and the ground as horizontal, find the range of the ball and its initial direction of motion.

15 An aeroplane is climbing at an angle of 2° while maintaining a speed of 400 m s^{-1}. A package is released and travels a horizontal distance of 2500 m before hitting the ground. By modelling the package as a particle projected with initial velocity the same as the velocity of the aeroplane, find the height of the aeroplane above the ground at the moment the package was released.

16 A child throws a ball with a speed of 20 m s^{-1} from a window 5 m above horizontal ground. If the ball hits the ground 1.5 seconds later find the direction of projection.

17 A stone is projected from a point O on a cliff with a speed of 20 m s^{-1} at an angle of elevation of 30°. T seconds later the angle of depression of the stone from O is 45°. Find the value of T.

18 A particle is projected from a point with position vector $(3\mathbf{i} + 3\mathbf{j})$ m relative to a fixed origin O where $\mathbf{i}$ and $\mathbf{j}$ are unit vectors in the horizontal and upward vertical directions respectively. Two seconds later the particle passes through the point with position vector $(13\mathbf{i} + 8\mathbf{j})$ m. Find its initial velocity vector.

19 A particle is projected with velocity vector $(10\mathbf{i} + 15\mathbf{j})$ m s^{-1} where $\mathbf{i}$ and $\mathbf{j}$ are unit vectors in the horizontal and upward vertical directions respectively. Find its velocity vector 2 seconds later and its distance from the point of projection at this time.

20 A particle is projected from a point on level ground. Its maximum height is 20 m and it hits the ground 200 m from its point of projection. Find the time of flight and the angle of projection.

21 A ball thrown at an angle $\arcsin \frac{4}{5}$ to the horizontal just clears the top of a wall 60 m away horizontally 4 s after projection.

By modelling the ball as a particle find (a) the velocity of projection (b) the height of the wall (c) the distance beyond the wall at which the ball hits the ground for the first time.

22 A particle is projected from O with a speed of $50\,\mathrm{m\,s^{-1}}$ at an angle of $30°$ above the horizontal. Find the greatest height reached above the horizontal plane through O. Find the length of time for which the particle is more than $20\,\mathrm{m}$ above the plane. The particle reaches the horizontal plane through O again at P. Find the distance OP.

SUMMARY OF KEY POINTS

1 Constant acceleration
For a particle moving with constant acceleration:

$$v = u + at$$
$$s = \left(\frac{u+v}{2}\right)t$$
$$s = ut + \tfrac{1}{2}at^2$$
$$v^2 = u^2 + 2as$$

2 Speed-time graphs
The area under a speed-time graph in a given time interval is the distance travelled during that time.

The gradient of a speed-time graph is the acceleration of the moving particle.

3 Variable acceleration
For variable acceleration the relationships between the displacement x, velocity v and acceleration a of a particle are shown by:

Differentiate w.r.t. t x → v → a Integrate w.r.t. t

4 Projectiles
The horizontal speed of a projectile is unchanged throughout the motion. The vertical motion is subject to an acceleration of magnitude $g = 9.8\,\mathrm{m\,s^{-2}}$ vertically downwards. The four uniform acceleration equations apply.

Review exercise

1

Whenever a numerical value of g is required, take $g = 9.8\,\text{m s}^{-2}$.

1   *ABCDEF* is a regular hexagon with centre *O*. The vectors $\overrightarrow{OA}$ and $\overrightarrow{OB}$ are denoted by **a** and **b** respectively.

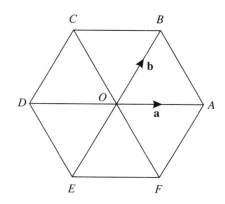

   (a) Find in terms of **a** and **b**:
      (i) $\overrightarrow{DO}$ (ii) $\overrightarrow{OE}$ (iii) $\overrightarrow{AB}$ (iv) $\overrightarrow{BC}$
      (v) $\overrightarrow{FC}$

   (b) The vector **b** is equal in magnitude and direction to $\overrightarrow{FA}$ and to $\overrightarrow{DC}$. Using only letters *O*, *A*, *B*, *C*, *D*, *E*, *F* write down vectors equal to:
      (i) $-$**a**   (ii) 2**a**   (iii) **a** $-$ **b**   (iv) **b** $-$ **a**
      (v) 2(**a** $-$ **b**).

2   *ABCDE* is a pentagon with $\overrightarrow{AB} = $ **p**, $\overrightarrow{BC} = $ **q**, $\overrightarrow{AE} = $ **r** and $\overrightarrow{ED} = $**s**.

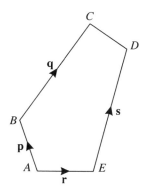

   (a) Express in terms of **p**, **q**, **r** and **s**:
      (i) $\overrightarrow{BE}$ (ii) $\overrightarrow{CD}$ (iii) $\overrightarrow{EC}$ (iv) $\overrightarrow{AD}$
      (v) $\overrightarrow{BD}$.

   (b) Given *BE* is parallel to *CD* and twice its length, write down an equation in **p**, **q**, **r** and **s**.

   (c) Given *AD* is parallel to *BC*, write down an equation in **q**, **r**, **s** and a scalar *k*.

3   *OAB* is an equilateral triangle. The vectors $\overrightarrow{OA}$ and $\overrightarrow{OB}$ are denoted by **a** and **b** respectively. Copy the figure and mark points whose position vectors relative to *O* are:

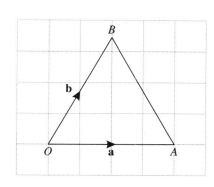

   (a) **a** $+$ **b**   (b) **a** $- \frac{1}{2}$**b**   (c) **a** $-$ 2**b**   (d) $-$**a** $+$ **b**
   (e) $-$**a** $- \frac{1}{2}$**b**.

4

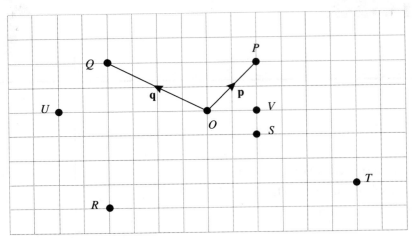

On the square grid shown, $\overrightarrow{OP} = \mathbf{p}$ and $\overrightarrow{OQ} = \mathbf{q}$. Express the following in terms of $\mathbf{p}$ or $\mathbf{q}$ or both $\mathbf{p}$ and $\mathbf{q}$.

(a) $\overrightarrow{OR}$ (b) $\overrightarrow{OS}$ (c) $\overrightarrow{OT}$ (d) $\overrightarrow{OU}$ (e) $\overrightarrow{OV}$

(f) $\overrightarrow{ST}$ (g) $\overrightarrow{UT}$ (h) $\overrightarrow{PT}$ (i) $\overrightarrow{QV}$ (j) $\overrightarrow{UP}$.

5

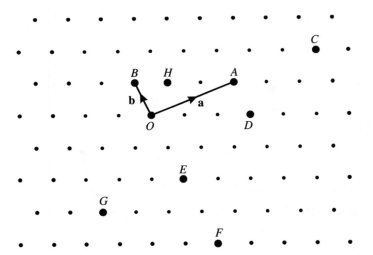

On the triangular grid shown, $\overrightarrow{OA} = \mathbf{a}$ and $\overrightarrow{OB} = \mathbf{b}$. Express the following in terms of $\mathbf{a}$ or $\mathbf{b}$ or both $\mathbf{a}$ and $\mathbf{b}$.

(a) $\overrightarrow{OC}$ (b) $\overrightarrow{OD}$ (c) $\overrightarrow{OE}$ (d) $\overrightarrow{OF}$ (e) $\overrightarrow{AE}$

(f) $\overrightarrow{OG}$ (g) $\overrightarrow{OH}$ (h) $\overrightarrow{ED}$ (i) $\overrightarrow{EC}$ (j) $\overrightarrow{FD}$.

6 At time $t = 0$, two ice skaters John (J) and Norma (N) have position vectors **40j** metres and **20i** metres relative to an origin 0 at the centre of an ice rink, where **i** and **j** are unit vectors perpendicular to each other. John has constant velocity $5\mathbf{i}\,\mathrm{m\,s^{-1}}$ and Norma has constant velocity $(3\mathbf{i} + 4\mathbf{j})\,\mathrm{m\,s^{-1}}$.

(a) Show that the skaters will collide and find the time at which the collision takes place.

(b) On another occasion, John has position vector **40j** metres. He wishes to skate in a straight line to the point with position vector **30i** metres. Given that his speed is constant at $5\,\mathrm{m\,s^{-1}}$, find his velocity.

7 In this question **i** and **j** are horizontal unit vectors and at right angles to each other.

At 12:00 a helicopter A sets out from its base O and flies with speed $120\,\mathrm{km\,h^{-1}}$ in the direction of the vector $3\mathbf{i} + 4\mathbf{j}$.

(a) Find the velocity vector of A.

At 12:20 that day another helicopter B sets out from O and flies with speed $150\,\mathrm{km\,h^{-1}}$ in the direction of the vector $24\mathbf{i} + 7\mathbf{j}$.

(b) Find the velocity vector of B.

(c) Find the position vectors of A and B at 13:00.

(d) Calculate the distance of A from B at 13:00.

At 13:30 B makes an emergency landing. A immediately changes direction and flies at $120\,\mathrm{km\,h^{-1}}$ in a straight line to B.

(e) Find the position vector of B from A at 13:30.

(f) Determine the time when A reaches B.

8 (a) Given that the vector $a\mathbf{i} + 3\mathbf{j}$ is parallel to the vector $2\mathbf{i} + 12\mathbf{j}$ find a.

(b) Given that the vector $3\mathbf{i} + b\mathbf{j}$ is parallel to $12\mathbf{i} - 20\mathbf{j}$ find b.

(c) Given that the vector $7\mathbf{i} + c\mathbf{j}$ has magnitude 25 find the two possible values of c.

9 The position vector **r** of a particle P at time t is given by $\mathbf{r} = t^2\mathbf{i} + (12 - t)\mathbf{j}$. Find the value of t when:

(a) **r** is parallel to the vector **i**

(b) **r** is parallel to the vector $\mathbf{i} + \mathbf{j}$

(c) the velocity **v** is parallel to the vector $\mathbf{i} - \mathbf{j}$.

10 At 11:00 hours the position vector of an aircraft relative to an airport O is $(200\mathbf{i} + 30\mathbf{j})$ km, $\mathbf{i}$ and $\mathbf{j}$ being unit vectors due east and due north respectively. The velocity of the aircraft is $(180\mathbf{i} - 120\mathbf{j})$ km h^{-1}.

Find (a) the time when the aircraft is due east of the airport O

(b) how far it then is from O

(c) how far it is from O at 12:00.

11 A train is uniformly retarded from 35 m s^{-1} to 21 m s^{-1} over a distance of 350 m. Calculate:

(a) the retardation

(b) the total time taken under this retardation to come to rest from a speed of 35 m s^{-1}. [L]

12 A car uniformly accelerates from rest at 0.7 m s^{-2} for 6 s. The car then immediately uniformly retards through a distance of 10.5 m and comes to rest. Calculate:

(a) the greatest speed of the car

(b) the time during which the car retards. [L]

13 A particle moves with uniform acceleration $\frac{1}{2}$ m s^{-2} in a horizontal line ABC. The speed of the particle at C is 80 m s^{-1} and the times taken from A to B and from B to C are 40 s and 30 s respectively. Calculate:

(a) the speed of the particle at A

(b) the distance BC. [L]

14 (a) A car, moving with uniform acceleration along a straight level road, passed points A and B when moving with speed 30 m s^{-1} and 60 m s^{-1} respectively. Find the speed of the car at the instant it passed C, the mid-point of AB. [L]

(b) State an assumption you have made about the car when forming the mathematical model you used to solve part (a).

15 A car is moving along a straight horizontal road at constant speed 18 m s^{-1}. At the instant when the car passes a lay-by, a motor-cyclist leaves the lay-by, starting from rest, and moves with constant acceleration 2.5 m s^{-2} in pursuit of the car. Given that the motor-cyclist overtakes the car T seconds after leaving the lay-by, calculate:

(a) the value of T

(b) the speed of the motor-cyclist at the instant of passing the car. [L]

16 A particle P moved in a straight line with constant retardation. At the instants when P passed through the points A, B and C it was moving with speeds $10\,\text{m}\,\text{s}^{-1}$, $7\,\text{m}\,\text{s}^{-1}$ and $3\,\text{m}\,\text{s}^{-1}$ respectively.

Prove that $\dfrac{AB}{BC} = \dfrac{51}{40}$.   [L]

17 (a) A ball is thrown vertically upwards at $7\,\text{m}\,\text{s}^{-1}$ from a point A which is $8\,\text{m}$ vertically above horizontal ground. Given that the ball moves freely under gravity, find:

    (i) the greatest height above the ground attained by the ball

    (ii) the speed of the ball, in $\text{m}\,\text{s}^{-1}$ to 3 significant figures, at the instant when it strikes the ground.   [L]

(b) State two assumptions you have made about the ball and the forces affecting its motion in solving part (a).

18 A stone is thrown vertically upwards with initial speed $28\,\text{m}\,\text{s}^{-1}$. Find the time taken to reach the greatest height that it attains above the point of projection, and find this height.   [L]

19 A ball is thrown vertically upwards and takes 3 seconds to reach its highest point. Find the times at which the ball is $39.2\,\text{m}$ above its point of projection.   [L]

20 A particle starts from rest at O and moves along Ox. During the first 4 seconds of its motion it accelerates uniformly to a speed of $12\,\text{m}\,\text{s}^{-1}$. The particle continues at this speed for 10 seconds. It then decelerates uniformly, coming to rest after a further T seconds. Display this information on a speed-time graph.

Given that the total distance travelled is $180\,\text{m}$, calculate T.   [L]

21

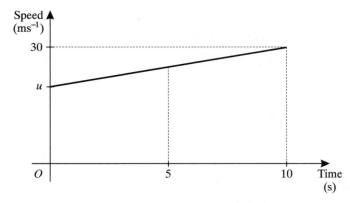

The diagram shows the speed-time graph for the motion of a car which is moving with constant acceleration in a straight line. The car passes a point A with speed $u\,\mathrm{m\,s^{-1}}$ and, 10 s later, it passes a point B with speed $30\,\mathrm{m\,s^{-1}}$.

(a) State, in terms of u, the speed of the car at the end of the first 5 seconds.

During the first 5 seconds the car travels 127.5 m.

(b) Calculate the value of u.

(c) Find, in $\mathrm{m\,s^{-2}}$, the magnitude of the acceleration of the car. [L]

22 A train starts from rest at a station, accelerates uniformly to its maximum speed of $15\,\mathrm{m\,s^{-1}}$, travels at this speed for a time, and then decelerates uniformly to rest at the next station. The distance from station to station is 1260 m, and the time spent travelling at maximum speed is three-quarters of the total journey time.

(a) Illustrate this information on a velocity-time graph.

(b) By using this graph, or otherwise, find the total journey time.

(c) Given also that the magnitude of the deceleration is twice the magnitude of the acceleration find, in $\mathrm{m\,s^{-2}}$, the magnitude of the acceleration. [L]

23 A particle A is constrained to move in a straight line. It starts from rest at X and accelerates at $2\,\mathrm{m\,s^{-2}}$ until it reaches speed V. It then travels with constant speed V for 60 s before decelerating at $1\,\mathrm{m\,s^{-2}}$ to come to rest at Y. The total time for the motion is 105 s. Find the distance XY.

A particle B starts from X at time $t = 0$ and moves along the same straight line towards Y with speed $(2t + 10)\,\mathrm{m\,s^{-1}}$, where t is the time in seconds. Calculate the time taken by B to reach Y from X.   [L]

24 A particle starting from rest at time $t = 0$ moves in a straight line and accelerates in the following way:

$$a = 3 \text{ when } 0 \leqslant t \leqslant 20$$
$$a = 1 \text{ when } 20 < t \leqslant 50$$
$$a = -2 \text{ when } 50 < t \leqslant 95$$

where a is the acceleration in $\mathrm{m\,s^{-2}}$ and t is the time in seconds. Find the speed of the particle when $t = 20$, $t = 50$ and $t = 95$. Sketch a time-speed graph for the particle in the interval $0 \leqslant t \leqslant 95$. Find the total distance travelled by the particle in this interval.   [L]

25 A particle moves along a straight line. It starts from rest, accelerates at $2\,\mathrm{m\,s^{-2}}$ for 3 seconds, and then decelerates at a constant rate, coming to rest in a further 8 seconds. Sketch a velocity-time graph for this motion. Find the total distance travelled by the particle during these 11 seconds.   [L]

26 (a) A cage goes down a vertical mine shaft 650 m deep in 42 s. During the first 16 s it is uniformly accelerated from rest to its maximum speed. For the next 10 s it moves at this maximum speed. It is then uniformly retarded to rest. By drawing a velocity-time graph, or otherwise, find, in $\mathrm{m\,s^{-1}}$, the maximum speed.   [L]

(b) State an assumption you have made about the cage when forming the mathematical model you used to solve (a).

27 A particle moves so that at time t seconds its position vector, $\mathbf{r}$ m, relative to a fixed origin is given by:

$$\mathbf{r} = (t^2 - 4t)\mathbf{i} + (t^3 + at^2)\mathbf{j}$$

where a is a constant.

(a) Find an expression for the velocity of the particle at time t seconds.

(b) Given that the particle comes to instantaneous rest, find the value of a.   [L]

28 Two particles, A and B, move in the plane of cartesian coordinate axes Ox, Oy. At time t seconds the position vectors of A and B, referred to the origin O, are $(t^2\mathbf{i} + 4t\mathbf{j})\,\mathrm{m}$ and $[2t\mathbf{i} + (t+1)\mathbf{j}]\,\mathrm{m}$ respectively.

(a) Prove that the particles never collide.

(b) Show that the velocity of B is constant, and calculate the magnitude and direction of this velocity.

(c) Find the value of t when A and B have parallel velocities, and find the distance between A and B at this instant.   [L]

29 A particle X, moving along a straight line with constant speed $4\,\mathrm{m\,s^{-1}}$, passes through a fixed point O. Two seconds later another particle Y, moving along the same straight line and in the same direction, passes through O with speed $6\,\mathrm{m\,s^{-1}}$. Given that Y is subject to a constant deceleration of magnitude $2\,\mathrm{m\,s^{-2}}$:

(a) state the velocity and displacement of each particle t seconds after Y passed through O

(b) find the shortest distance between the particles after they have both passed through O

(c) find the value of t when the distance between the particles has increased to $23\,\mathrm{m}$.   [L]

30 A stone is thrown from a point $12.6\,\mathrm{m}$ above horizontal ground with speed $7\,\mathrm{m\,s^{-1}}$ at an angle of $30°$ above the horizontal. Calculate the time which elapses before the stone hits the ground.   [L]

31 A bowler delivers a ball horizontally at $30\,\mathrm{m\,s^{-1}}$, his hand being $2\,\mathrm{m}$ above the horizontal ground at the instant that the ball is released. Find, in metres to the nearest metre, the horizontal distance that the ball travels before striking the ground.   [L]

32 At time $t = 0$ a ball is kicked from a point A on horizontal ground and moves freely under gravity. At time $t = 4\,\mathrm{s}$ the ball hits the ground at B, where $AB = 60\,\mathrm{m}$. Calculate the horizontal and vertical components of the velocity of the ball as it leaves A.   [L]

33 A ball was thrown from a balcony above a horizontal lawn. The velocity of projection was $10 \, \text{m s}^{-1}$ at an angle of elevation α, where $\tan \alpha = \frac{3}{5}$. The ball moved freely under gravity and took $3 \, \text{s}$ to reach the lawn from the instant when it was thrown. Calculate:

(a) the vertical height above the lawn from which the ball was thrown

(b) the horizontal distance between the point of projection and the point A at which the ball hit the lawn

(c) the angle, to the nearest degree, between the direction of the velocity of the ball and the horizontal at the instant when the ball reached A.   [L]

34 The unit vectors $\mathbf{i}$ and $\mathbf{j}$ are horizontal and vertically upwards respectively. A particle is projected with velocity $(8\mathbf{i} + 10\mathbf{j}) \, \text{m s}^{-1}$ from a point O at the top of a cliff and moves freely under gravity. Six seconds after projection, the particle strikes the sea at the point S. Calculate: [L]

(a) the horizontal distance between O and S

(b) the vertical distance between O and S.

(c) At time T seconds after projection, the particle is moving with velocity $(8\mathbf{i} - 14.5\mathbf{j}) \, \text{m s}^{-1}$. Find the value of T and the position vector, relative to O, of the particle at this instant.   [L]

35 A golf ball is hit from a point O on horizontal ground and moves freely under gravity. The horizontal and vertical components of the initial velocity of the ball are $2u \, \text{m s}^{-1}$ and $u \, \text{m s}^{-1}$ respectively. The ball hits the horizontal ground at a point whose distance from O is $80 \, \text{m}$.

(a) Show that $u = 14$.

(b) Show that the highest point of the path of the ball is $10 \, \text{m}$ above the ground.

(c) Find, in m s^{-1} to 2 significant figures, the speed of the ball $2\frac{1}{2}$ seconds after the ball has been hit.

(d) State four assumptions you have made about the ball and the forces acting on it during its flight.   [L]

36 A pilot, flying an aeroplane in a straight line at a constant speed of $196\,\mathrm{m\,s^{-1}}$ and at a constant height of $2000\,\mathrm{m}$, drops a bomb on a stationary ship in the vertical plane through the line of flight of the aeroplane. Assuming that the bomb falls freely under gravity, calculate:

(a) the time which elapses after release before the bomb hits the ship

(b) the horizontal distance between the aeroplane and the ship at the time of release of the bomb

(c) the speed of the bomb just before it hits the ship.   [L]

37 A particle P is projected, from a point O on horizontal ground, with speed $V\,\mathrm{m\,s^{-1}}$ and angle of elevation α. The particle moves freely under gravity. After 5 seconds the components of velocity of P are $20\,\mathrm{m\,s^{-1}}$ horizontally and $14\,\mathrm{m\,s^{-1}}$ vertically *downwards*.

(a) Show that $\tan\alpha = \frac{7}{4}$.

(b) Calculate, to 1 decimal place, the value of V.

(c) Find the greatest height above the ground reached by P.

(d) Calculate the speed of P, in $\mathrm{m\,s^{-1}}$ to 1 decimal place, 7 seconds after leaving O.   [L]

Statics of a particle

4

'Static' means 'stationary' or 'at rest', and this chapter looks at the forces acting on a particle that is not moving. Later in this chapter the various types of forces that may act on a particle will be considered. At this point you only need to remember from chapter 2 that you can describe a force acting on a particle by stating its magnitude and its direction.

Forces are measured in **newtons**. The precise definition of a newton is given in chapter 5 (p. 121). A force of magnitude 8 newtons is usually written 8 N.

4.1 Resultant forces

Suppose two forces $\mathbf{F}_1$ and $\mathbf{F}_2$ act on a particle as shown. As forces are vectors they may be added by the triangle law (see p. 17) You can redraw the diagram as:

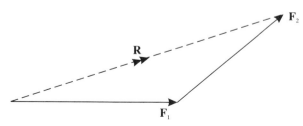
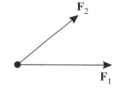

Then $\mathbf{R} = \mathbf{F}_1 + \mathbf{F}_2$ is the **resultant** of the two forces. Use a double arrow to indicate that $\mathbf{R}$ is a resultant force rather than a third force acting on the particle.

Another way of adding forces is to use the **parallelogram rule**. (See p. 17).

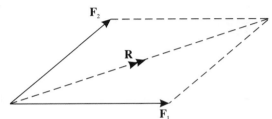

The resultant **R** is then given by the diagonal of the parallelogram. The advantage of this method is that it shows the two forces and their resultant acting at the same point. However, it is usually more convenient in calculations to use the triangle law.

Example 1

Two forces of magnitude 5 N and 6 N act on a particle. They act at right angles. Find the magnitude and direction of their resultant.

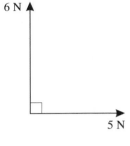

Let R N be the magnitude of the resultant and θ the angle it makes with the 5 N force. The vector triangle is:

By Pythagoras' Theorem:
$$R^2 = 6^2 + 5^2 = 61$$
$$R = 7.81$$

Also:
$$\tan\theta = \frac{6}{5} = 1.2$$
$$\theta = 50.2°$$

The resultant is a force of magnitude 7.81 N at an angle of 50.2° to the 5 N force.

Example 2

Two forces acting on a particle have magnitudes 8 N and 3 N. The angle between their directions is 60°. Find the resultant force acting on the particle.

The force diagram is:

The vector triangle for this situation is:

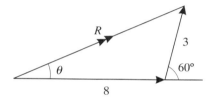

Let the resultant be of magnitude R N and make an angle of θ with the 8 N force.

By the cosine rule (see p. 2):

$$R^2 = 3^2 + 8^2 - 2 \times 8 \times 3 \times \cos 120°$$
$$= 9 + 64 - 48(-\tfrac{1}{2})$$
$$= 97$$
$$R = \sqrt{97} = 9.85$$

By the sine rule (see p. 1):

$$\frac{R}{\sin 120°} = \frac{3}{\sin \theta}$$

Giving:
$$\sin \theta = \frac{3 \sin 120°}{R}$$
$$= \frac{3 \sin 120°}{\sqrt{97}}$$

So:
$$\theta = 15.3°$$

The resultant force is of magnitude 9.85 N at an angle of 15.3° to the 8 N force.

(Notice that to achieve maximum accuracy in θ you should use $\sqrt{97}$ for R and not the corrected value of 9.85.)

Finding the resultant of more than two forces

The resultant $\mathbf{R}_1$ of any two forces $\mathbf{P}$ and $\mathbf{Q}$ acting on a particle can be found by constructing a vector triangle.

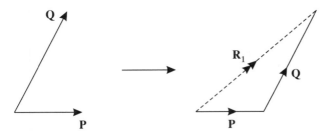

This method can be extended to any number of forces. Suppose a third force **S** also acts on the particle. The resultant $\mathbf{R}_2$ of the three forces is then the resultant of $\mathbf{R}_1$ and **S**. You can find $\mathbf{R}_2$ by drawing a **polygon of forces**: the resultant $\mathbf{R}_2$ is represented by the line required to complete the polygon.

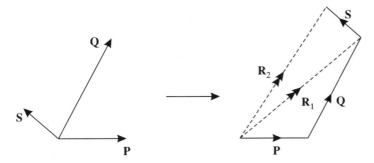

You can generalise this method to any number of forces. The magnitude and direction of the resultant can be calculated using trigonometry.

However, it is easier to find the resultant of a set of forces if the forces are expressed in terms of cartesian components or resolutes (see section 2.3) In the following example the forces are given in the **i, j** notation.

Example 3

Three forces $(2\mathbf{i} - \mathbf{j})\,\text{N}$, $3\mathbf{i}\,\text{N}$ and $(-\mathbf{i} + 4\mathbf{j})\,\text{N}$, where **i** and **j** are unit vectors due east and due north respectively, act on a particle. Find the magnitude and direction of the resulting force.

The forces are given in the **i, j** notation, so you can add the vectors by the method given in section 2.3 (p. 22) to obtain the resultant **R** N:

$$\mathbf{R} = (2\mathbf{i} - \mathbf{j}) + 3\mathbf{i} + (-\mathbf{i} + 4\mathbf{j})$$
$$= 4\mathbf{i} + 3\mathbf{j}$$

The magnitude of the resultant is:

$$|\mathbf{R}| = \sqrt{(4^2 + 3^2)} = \sqrt{25} = 5$$

The diagram shows that:

$$\tan \theta = \tfrac{3}{4}$$
$$\theta = 36.9°$$

The resultant is therefore of magnitude 5 N and acts at an angle of 36.9° to unit vector **i**.

Exercise 4A

1 Forces **P** and **Q** act on a particle at the origin O along the coordinates axes Ox, Oy respectively. Calculate the magnitude of the resultant force and the angle it makes with the x axis when:

(a) $P = 4\,\text{N}$, $Q = 3\,\text{N}$
(b) $P = 5\,\text{N}$, $Q = 8\,\text{N}$
(c) $P = 8\,\text{N}$, $Q = 5\,\text{N}$
(d) $P = 24\,\text{N}$, $Q = 15\,\text{N}$
(e) $P = 9\,\text{N}$, $Q = 40\,\text{N}$.

2 Two forces **P** and **Q** act on a particle at the origin O. Force **P** acts along Ox, force **Q** makes an angle θ with Ox. Calculate the magnitude of the resultant force and the angle it makes with Ox when:

(a) $P = 4\,\text{N}$, $Q = 3\,\text{N}$, $\theta = 60°$
(b) $P = 3\,\text{N}$, $Q = 4\,\text{N}$, $\theta = 60°$
(c) $P = 5\,\text{N}$, $Q = 5\,\text{N}$, $\theta = 60°$
(d) $P = 10\,\text{N}$, $Q = 12\,\text{N}$, $\theta = 120°$
(e) $P = 8\,\text{N}$, $Q = 5\,\text{N}$, $\theta = 120°$.

3 Forces **P**, **Q** and **R** act on a particle at O in the plane of the coordinate axes Ox, Oy. Force **P** acts along Ox, **Q** acts along Oy, and **R** acts at an angle θ with Ox, in the first quadrant. Calculate the magnitude of the resultant force and the angle it makes with Ox when:

(a) $P = 3\,\text{N}$, $Q = 4\,\text{N}$, $R = 5\,\text{N}$, $\theta = 60°$
(b) $P = 4\,\text{N}$, $Q = 3\,\text{N}$, $R = 5\,\text{N}$, $\theta = 60°$
(c) $P = 5\,\text{N}$, $Q = 6\,\text{N}$, $R = 7\,\text{N}$, $\theta = 45°$
(d) $P = 6\,\text{N}$, $Q = 6\,\text{N}$, $R = 6\,\text{N}$, $\theta = 60°$
(e) $P = 6\,\text{N}$, $Q = 6\,\text{N}$, $R = 6\,\text{N}$, $\theta = 45°$.

4 Forces **P**, **Q** and **R** act on a particle at the origin O. Find the resultant of the three forces in terms of **i** and **j**, the unit vectors along Ox and Oy respectively, when:

   (a) **P** = 2**i** + 5**j**, **Q** = 3**i** + 4**j**, **R** = 5**i** + 2**j**

   (b) **P** = 2**i** + 5**j**, **Q** = 3**i**, **R** = 5**j**

   (c) **P** = 2**i** + 5**j**, **Q** = 2**i** − 5**j**, **R** = 3**i**

   (d) **P** = 2**i** + 5**j**, **Q** = **i** − 7**j**, **R** = −3**i** − 4**j**

   (e) **P** = 2**i** + 5**j**, **Q** = −3**i** − 4**j**, **R** = **i** − **j**.

5 For each part of question 4 find the magnitude of the resultant force and the angle it makes with Ox.

6 Forces **P**, **Q**, **R** and **S** act on a particle at O in the plane of the coordinate axes Ox, Oy, making angles p, q, r, s respectively with Ox, each angle being measured in the anticlockwise sense. By drawing a polygon of forces find the magnitude of their resultant and the angle it makes with Ox when:

   (a) $P = 2\,\text{N}$, $Q = 3\,\text{N}$, $R = 4\,\text{N}$, $S = 5\,\text{N}$, $p = 0°$, $q = 40°$,
      $r = 100°$, $s = 150°$

   (b) $P = 3\,\text{N}$, $Q = 3\,\text{N}$, $R = 5\,\text{N}$, $S = 5\,\text{N}$, $p = 0°$, $q = 40°$,
      $r = 130°$, $s = 220°$

   (c) $P = 1\,\text{N}$, $Q = 2\,\text{N}$, $R = 3\,\text{N}$, $S = 4\,\text{N}$, $p = 10°$, $q = 70°$,
      $r = 100°$, $s = 300°$.

4.2 Resolving forces into components

You saw in chapter 2 that it is often useful to resolve a vector into **components** or **resolutes**. This is a very important operation when the vectors you are dealing with are forces. In this case the components are often called the **resolved parts** of the force.

In chapter 2 only components in the direction of the coordinate axes were considered.

But when you are dealing with forces it is sometimes useful to find components horizontally and vertically, while at other times it is more useful to find the components parallel to and at right angles to a given plane. You should choose the directions for the components that produce the simplest possible equations. This is determined by the direction of forces such as the weight of the particle and the frictional force.

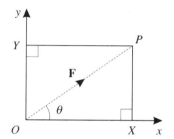

Suppose the force **F** acts at an angle θ to the x-axis.
By the triangle law:

$$\overrightarrow{OP} = \overrightarrow{OX} + \overrightarrow{XP}$$
$$= \overrightarrow{OX} + \overrightarrow{OY}$$

Then the components or resolutes of **F** along the x- and y-axes are $\overrightarrow{OX}$ and $\overrightarrow{OY}$.

By trigonometry:

$$\frac{OX}{OP} = \cos\theta$$

So:
$$OX = OP\cos\theta = F\cos\theta$$

And:
$$\frac{OY}{OP} = \sin\theta$$

So:
$$OY = OP\sin\theta = F\sin\theta$$

So the components of **F** along the x- and y-axes are:

$$F\cos\theta \qquad \text{and} \qquad F\sin\theta$$

and:
$$\mathbf{F} = F\cos\theta\,\mathbf{i} + F\sin\theta\,\mathbf{j}$$

As $\sin\theta = \cos(90° - \theta)$ you can write this as:

$$\mathbf{F} = F\cos\theta\,\mathbf{i} + F\cos(90° - \theta)\mathbf{j}$$

F makes angles θ and $(90° - \theta)$ with the x- and y-axes respectively. This leads to a general rule for obtaining components or resolutes of a force:

■ **The component of a force in any direction is the product of the magnitude of the force and the cosine of the angle between the force and the required direction.**

Notice in particular that the component of a force in a direction perpendicular to that force is zero (as $\theta = 0°$ in that case and $\cos(90-0)° = \cos 90° = 0$).

Example 4

A force of magnitude 5 N acts at an angle of 62° to the horizontal. Find the magnitudes of the horizontal and vertical components of this force.

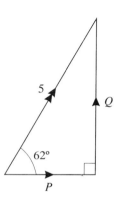

Let P and Q be the horizontal and vertical components of the force.

$$P = 5\cos 62° = 2.35$$

$$Q = 5\cos(90-62)°$$

Or more usually:

$$Q = 5 \sin 62° = 4.14$$

So the magnitudes of the horizontal and vertical components are 2.35 N and 4.14 N.

Example 5

A particle lies on the face of a plane inclined at an angle θ to the horizontal. It is acted on by a force **R** perpendicular to the plane and a force **W** vertically downwards. Find the components of the resultant force on the particle:

(a) horizontally and vertically
(b) down the plane and perpendicular to the plane.

The force diagram for this situation is:

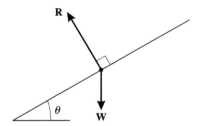

Before attempting to obtain components it is necesary to work out some important angles. The diagram below shows the essentials.

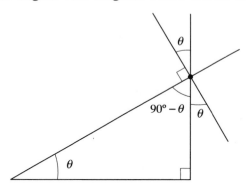

(a) Resolve the force **R** into horizontal and vertical components:

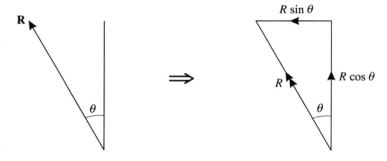

The force **W** is vertical, so it has *no* horizontal component. So:

horizontal component of the resultant is $R \sin \theta \leftarrow$
vertical component of the resultant is $R \cos \theta - W \uparrow$

(b) Clearly **R** will not have no component down the plane as **R** is perpendicular to this direction.

Resolving **W** in these directions gives:

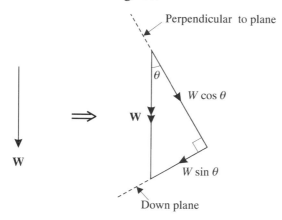

So:

component of resultant down plane is $W \sin \theta \nearrow$
component of resultant perpendicular to plane is $R - W \cos \theta \searrow$

The arrows are added to make it quite clear in which direction each component force acts.

When you combine the components to find the resultant vector, you must be careful to show the components in the correct direction in your force diagram.

4.3 Finding the resultant of several forces by resolving them into components

Suppose several forces act on a particle. One way to work out the resultant of the forces is to find the components of each force in two perpendicular directions. You can then find the algebraic sum of these components to give the magnitude and direction of the resultant.

Example 6

Here are four forces acting on a particle. Find the magnitude and direction of the resultant force.

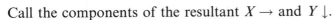

The question does not tell you which directions you should choose when you resolve the forces into components. Take → along the 5 N force and ↓ along the 30 N force. These directions are perpendicular. The 30 N force will not contribute anything to the horizontal component, and the 5 N force will not contribute to the vertical component.

Call the components of the resultant $X \rightarrow$ and $Y \downarrow$.

Resolving horizontally → gives: $X = 5 + 10 \cos 60° - 8 \cos 50°$
$$= 4.86$$

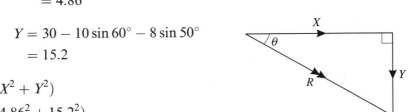

Resolving vertically ↓ gives: $Y = 30 - 10 \sin 60° - 8 \sin 50°$
$$= 15.2$$

So:
$$R = \sqrt{(X^2 + Y^2)}$$
$$= \sqrt{(4.86^2 + 15.2^2)}$$
$$= 16.0$$

and:
$$\tan \theta = \frac{Y}{X} = \frac{15.2}{4.86}$$
$$\theta = 72.3°$$

So the resultant force is of magnitude 16.0 N and makes an angle of 72.3° with the 5 N force.

Exercise 4B

1 A force **F** acts on a particle at the origin O in the plane of the coordinate axes Ox, Oy, at an angle θ to Ox, θ being measured in the anticlockwise sense. Find the components of **F** along Ox and Oy when:
(a) $F = 4\,\text{N}$, $\theta = 30°$
(b) $F = 6\,\text{N}$, $\theta = 60°$
(c) $F = 6\,\text{N}$, $\theta = 120°$
(d) $F = 8\,\text{N}$, $\theta = 150°$
(e) $F = 10\,\text{N}$, $\theta = 240°$.

2 A force **F** acts on a particle at an angle of θ to the horizontal. Find the horizontal and vertical components of **F** when:

(a) $F = 2\,\text{N}, \theta = 20°$

(b) $F = 10\,\text{N}, \theta = 30°$

(c) $F = 20\,\text{N}, \theta = 50°$

3 A force **W** acts vertically downwards on a particle at rest on a plane inclined at an angle θ to the horizontal. Find the components of **W** along and perpendicular to the plane when:

(a) $W = 2\,\text{N}, \theta = 30°$

(b) $W = 4\,\text{N}, \theta = 50°$

(c) $W = 10\,\text{N}, \theta = 75°$.

4

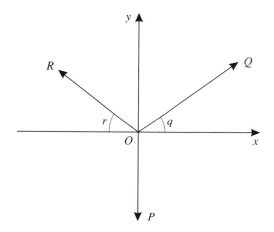

Forces **P**, **Q** and **R** act on a particle at rest at O, as shown in the diagram. By first finding the components of each force along Ox and Oy, find the magnitude of the resultant force and the angle it makes with Ox when:

(a) $P = 5\,\text{N}, Q = 4\,\text{N}, R = 3\,\text{N}, q = 20°, r = 40°$

(b) $P = 5\,\text{N}, Q = 6\,\text{N}, R = 4\,\text{N}, q = 20°, r = 40°$

(c) $P = 10\,\text{N}, Q = 4\,\text{N}, R = 5\,\text{N}, q = 30°, r = 60°$.

5

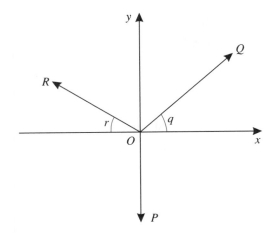

Forces P, Q and R act on a particle at rest at the origin O, as shown in the diagram. By first finding the components of each force along the line of action of Q, find the magnitude of the resultant force and the angle its line of action makes with the line of action of Q when:

(a) $P = 2\,\text{N}$, $Q = 3\,\text{N}$, $R = 4\,\text{N}$, $q = 20°$, $r = 70°$

(b) $P = 5\,\text{N}$, $Q = 6\,\text{N}$, $R = 7\,\text{N}$, $q = 30°$, $r = 60°$

(b) $P = 10\,\text{N}$, $Q = 20\,\text{N}$, $R = 30\,\text{N}$, $q = 60°$, $r = 30°$.

4.4 Equilibrium of coplanar forces

If all the forces acting on a particle are in the same plane they are called **coplanar forces**. This book deals only with coplanar forces. In the real world, of course, forces can act in all directions on a particle. If all the forces acting on a particle cancel each other out, so that nothing happens at all, the forces are said to be in **equilibrium**.

■ **A system of forces acting on a particle is said to be in equilibrium if their resultant is the zero vector.**

When forces in equilibrium are resolved into components in two fixed directions, the algebraic sum of the components in each direction will also be zero. This is the basic fact you need to know to deal with particles in equilibrium.

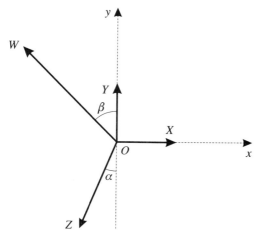

Consider a particle at O in equilibrium under the action of the forces **X**, **Y**, **W** and **Z** as shown above.

Resolving horizontally $\rightarrow$ gives: $X - W \sin \beta - Z \sin \alpha$

Resolving vertically $\uparrow$ gives: $Y + W \cos \beta - Z \cos \alpha$

As the particle is in equilibrium, each of these expressions must be equal to zero.

So: $X - W \sin \beta - Z \sin \alpha = 0$

and: $Y + W \cos \beta - Z \cos \alpha = 0$

There are two equations, so you can use them to find two unknown quantities (see Book P1, section 2.2). Whichever two directions you choose, you will always get two equations like these.

Example 7

A particle is in equilibrium under the action of the forces shown. Find the magnitude of the forces **P** and **S**.

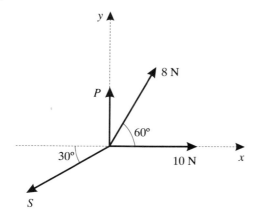

Resolving in the $0x$ direction $\rightarrow$ gives: $10 + 8 \cos 60° - S \cos 30° = 0$

As $\cos 60° = \frac{1}{2}$ and $\cos 30° = \frac{\sqrt{3}}{2}$ (see Book P1, p. 142):

$$10 + 8 \times \tfrac{1}{2} - S \times \tfrac{\sqrt{3}}{2} = 0$$
$$14 - S \times \tfrac{\sqrt{3}}{2} = 0$$
$$S = \frac{28}{\sqrt{3}} = \frac{28\sqrt{3}}{3}$$

Resolving in $0y$ direction $\uparrow$ gives: $\quad P + 8 \sin 60° - S \sin 30° = 0$

As $\sin 60° = \frac{\sqrt{3}}{2}$ and $\sin 30° = \frac{1}{2}$ (see Book P1, p.142):

$$P + 8 \times \tfrac{\sqrt{3}}{2} - S \times \tfrac{1}{2} = 0$$
$$P = \tfrac{1}{2}S - 4\sqrt{3}$$

Substitute the value calculated above for S:

$$P = \frac{14\sqrt{3}}{3} - 4\sqrt{3}$$
$$= \frac{2\sqrt{3}}{3}$$

So the force **P** has magnitude $\frac{2\sqrt{3}}{3}$ N (1.15 N) and the force **S** has magnitude $\frac{28\sqrt{3}}{3}$ N (16.2 N).

Example 8

A particle is in equilibrium under the forces $(6\mathbf{i} + 4\mathbf{j})$ N, $(-2\mathbf{i} - 5\mathbf{j})$ N and $(a\mathbf{i} + b\mathbf{j})$ N. Find the values of a and b.

As the particle is in equilibrium:

$$(6\mathbf{i} + 4\mathbf{j}) + (-2\mathbf{i} - 5\mathbf{j}) + (a\mathbf{i} + b\mathbf{j}) = \mathbf{0}$$
$$\mathbf{i}(6 - 2 + a) + \mathbf{j}(4 - 5 + b) = \mathbf{0}$$

As the resultant is the zero vector, the coefficient of **i** is zero. So:

$$6 - 2 + a = 0$$
$$a = -4$$

The coefficient of **j** is also zero. So:

$$4 - 5 + b = 0$$
$$b = 1$$

Exercise 4C

1

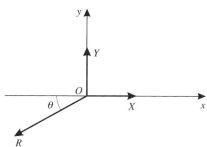

A particle is in equilibrium at O under the action of forces **X**, **Y** and **R**, as shown in the diagram. Find the magnitude of **R** and the angle θ when:

(a) $X = 3$ N, $Y = 4$ N

(b) $X = 5$ N, $Y = 8$ N

(c) $X = 5$ N, $Y = 12$ N.

2

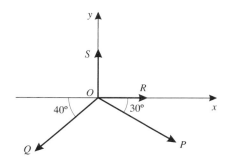

A particle is in equilibrium at O under the action of forces **P**, **Q**, **R** and **S**, as shown in the diagram. Find the magnitudes of forces R and S when:

(a) $P = 5$ N, $Q = 6$ N

(b) $P = 8$ N, $Q = 10$ N

(c) $P = 12$ N, $Q = 20$ N.

3 A particle is in equilibrium at O under the action of forces **P** N, **Q** N and a third force $(x\mathbf{i} + y\mathbf{j})$ N. Find x and y when:

(a) $\mathbf{P} = 2\mathbf{i} + 3\mathbf{j}$, $\mathbf{Q} = 3\mathbf{i} + 5\mathbf{j}$

(b) $\mathbf{P} = 2\mathbf{i} - 3\mathbf{j}$, $\mathbf{Q} = -3\mathbf{i} - 4\mathbf{j}$

(c) $\mathbf{P} = 3\mathbf{i} - \mathbf{j}$, $\mathbf{Q} = 2\mathbf{i} - 5\mathbf{j}$

(d) $\mathbf{P} = -4\mathbf{i} + 5\mathbf{j}$, $\mathbf{Q} = -\mathbf{i} - 7\mathbf{j}$

(e) $\mathbf{P} = -\mathbf{j}$, $\mathbf{Q} = \mathbf{i} + 2\mathbf{j}$.

4.5 Types of force

So far the forces acting on a particle have been examined, without considering the *nature* of these forces. Some of the mechanical forces that may act on a particle are considered in this section.

Weight

All particles falling freely under gravity have the same acceleration. This constant acceleration is denoted by the symbol g. The acceleration must be caused by a force acting on the particle. This force is called the **weight** of the particle. If the mass of the particle is m kg then (as shown in section 5.1):

$$\text{weight} = m\text{g}$$

where the weight is a force measured in newtons.

Throughout this book the value of g will be taken as $9.8 \,\text{m s}^{-2}$, although it has different values at different points on the surface of the Earth. Weight is an example of a **non-contact force**: an object does not have to be in contact with the Earth for the force of gravity to act on it. All the other forces considered below are **contact forces**.

Tension

Imagine a particle of mass m kg hanging in equilibrium at the end of a string. As the particle is in equilibrium (it doesn't fall or fly upwards) the weight mg acting downwards must be balanced by an equal and opposite force acting upwards. This force is called the **tension** T in the string.

Thrust

Now imagine a similar situation where a particle is supported by a vertical spring from below. As the particle is in equilibrium there must be an upward force in the spring to balance the downward force of weight. This is called the **thrust**.

4.6 Resolving a contact force into normal and frictional components

Think of a particle at rest on a horizontal table. If the table is removed the particle will fall to the ground. There must therefore be an upward force, supplied by the table, which balances the weight of the particle. This is called the **normal reaction** or **normal contact force** and is usually denoted by **R**.

If you try to push the particle across the table then in general the particle will not move. An additional **frictional force F** acts to balance the pushing force. The frictional force acts parallel to the table and *opposes* the motion that the pushing force is trying to create. Here is the force diagram of this situation:

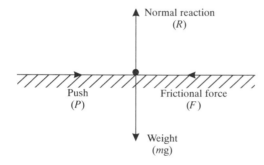

If the table has a perfectly *smooth* surface then $F = 0$ and there is only a normal reaction.

When you have a problem in which some or all of the above forces act on a particle, it is important to begin by drawing a clear diagram of reasonable size on which the forces are clearly indicated.

Example 9
Draw diagrams to show the forces acting on a particle in each of the following.

(a) A particle at rest on a *rough* inclined plane: **R** is perpendicular to the plane. **F** is up the plane as it opposes the motion which the weight of the particle on its own would produce.

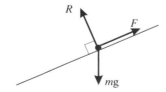

(b) A particle pulled across a *rough* horizontal table by a string inclined at 60° to the horizontal:

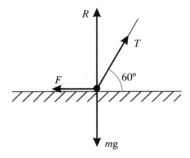

F opposes the force produced by the pull on the string.

(c) A particle held at rest on a *smooth* plane inclined at 30° to the horizontal, by a string parallel to the plane:

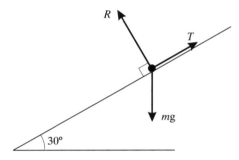

There is no frictional force in this case as the plane is smooth.

(d) A particle suspended from a horizontal beam by two unequal strings:

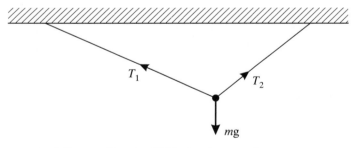

Some examples of equilibrium under these forces

Example 10

A particle of mass 5 kg is suspended in equilibrium by two light inextensible strings which make angles of 30° and 45° respectively with the horizontal. Find the tensions in the strings.

You will remember that 'inextensible' means the strings do not stretch.

The weight of the particle is 5g N (see p. 106). The force diagram is:

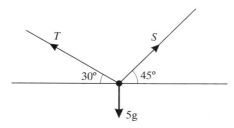

As the particle is in equilibrium, the components of the forces in each direction must add up to zero.

Resolving horizontally $\rightarrow$ gives: $S \cos 45° - T \cos 30° = 0$

$$T = \frac{S \cos 45°}{\cos 30°}$$

You may wish to either keep the equation in trigonometrical form or you could use $\cos 45° = \frac{\sqrt{2}}{2}$ and $\cos 30° = \frac{\sqrt{3}}{2}$. (In the latter case you would get $T = S\sqrt{\frac{2}{3}}$.)

Resolving vertically $\uparrow$ gives: $\quad T \sin 30° + S \sin 45° - 5g = 0$

Now substitute the value of T found above:

$$S \cos 45° \times \left(\frac{\sin 30°}{\cos 30°}\right) + S \sin 45° = 5g$$

$$S \cos 45° \tan 30° + S \sin 45° = 5g$$

(or $S\frac{\sqrt{2}}{2}\left[\frac{1}{\sqrt{3}} + 1\right] = 5g$)

Using g $= 9.8$ gives:

$$S = 43.9 \, \text{N}$$

and: $\qquad\qquad\qquad T = 35.9 \, \text{N}$

Example 11

A particle of mass m kg rests in equilibrium on a rough plane inclined at 30° to the horizontal. Find the normal contact force and the frictional force in terms of m and g.

The force diagram is:

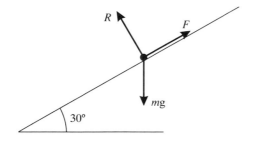

So that essentially:

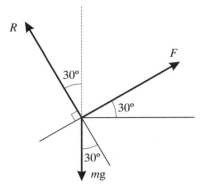

You can solve this problem in two ways, by choosing two different sets of directions in which to resolve the forces.

(i)   Take components along and perpendicular to the plane. **F** is along the plane and **R** is perpendicular to the plane. You only need, therefore, to resolve the weight into components along and perpendicular to the plane.

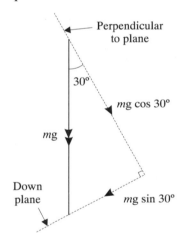

Resolving up the plane ↗ gives: $F - mg \sin 30° = 0$

As $\sin 30° = \frac{1}{2}$: $F = \dfrac{mg}{2}$

Resolving perpendicular to the plane ↘ gives: $R - mg \cos 30° = 0$

As $\cos 30° = \frac{\sqrt{3}}{2}$: $R = \dfrac{mg\sqrt{3}}{2}$

(ii)   Take components horizontally and vertically, using sin 30° and cos 30° as above:

Resolving vertically ↑ gives: $\quad R \cos 30° + F \sin 30° = mg$

$$R \times \frac{\sqrt{3}}{2} + F \times \frac{1}{2} = mg$$

Resolving horizontally → gives: $\quad F \cos 30° - R \sin 30° = 0$

So: $\quad\quad\quad\quad\quad\quad\quad\quad\quad\quad\quad\quad R = F\sqrt{3}$

Substitute for R in the other equation to get:

$$F \times \tfrac{3}{2} + F \times \tfrac{1}{2} = mg$$
$$F = \frac{mg}{2}$$

Hence: $\quad\quad\quad\quad\quad\quad\quad\quad\quad R = \frac{mg\sqrt{3}}{2}$

In the first method you obtained F and R directly. You may, however, find the equations easier to write down in the second method. Choose whichever method you find easier.

Exercise 4D

1 A small body is hanging in equilibrium at the end of a
   vertical string. Find
   (a) the weight of the body if its mass is $0.2\,\text{kg}$
   (b) the mass of the body if its weight is $4.9\,\text{N}$
   (c) the tension in the string if the mass of the body is $1\,\text{kg}$
   (d) the tension in the string if the weight of the body is $5\,\text{N}$.

2 A book is at rest in equilibrium on a horizontal table. Find
   (a) the normal reaction of the table on the book if the weight
       of the book is $8\,\text{N}$
   (b) the normal reaction of the table on the book if the mass
       of the book is $2.2\,\text{kg}$
   (c) the weight of the book if the normal reaction is $6\,\text{N}$
   (d) the mass of the book if the normal reaction is $10\,\text{N}$.

3

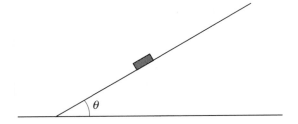

A particle of mass m kg is at rest on a rough plane inclined at an angle θ above the horizontal, as shown in the diagram. Find the frictional force exerted by the surface of the plane on the particle when:

(a) $m = 5$, $\theta = 30°$

(b) $m = 10$, $\theta = 40°$

(c) the weight of the particle is 8 N and $\theta = 20°$.

4

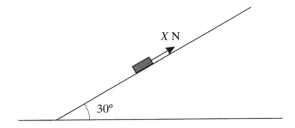

A particle of mass m kg is at rest on a rough plane inclined at an angle $30°$ above the horizontal. Force X N acts on the particle up the line of greatest slope of the plane, as shown in the diagram. Find the magnitude and direction of the frictional force when:

(a) $X = 10$, $m = 1$

(b) $X = 10$, $m = 2$

(c) $X = 10$, $m = 3$

(d) $X = 5$, $m = 2$

(e) $X = 5$ and the weight of the particle is 4 N.

5

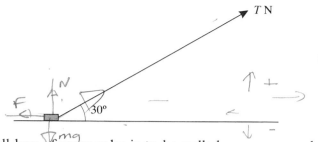

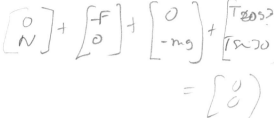

$$\begin{bmatrix} 0 \\ N \end{bmatrix} + \begin{bmatrix} -F \\ 0 \end{bmatrix} + \begin{bmatrix} 0 \\ -mg \end{bmatrix} + \begin{bmatrix} T\cos 3? \\ T\sin 30 \end{bmatrix}$$

$$= \begin{bmatrix} 0 \\ 0 \end{bmatrix}$$

A small box of mass m kg is to be pulled across a rough horizontal floor at a constant speed by a string inclined at $30°$ above the horizontal, as shown in the diagram. The tension in the string is T N. Model the box as a particle and find the frictional force F N exerted by the floor on the box and the normal reaction of the floor on the box when the box is on the point of sliding when:

$F = T\cos 30$

$N - mg + T\sin 30 = 0$

$F = \mu N$.

$\mu N = T\cos 30$

$N - mg + T\sin 30 = 0$

(a) $m = 5$, $T = 20$

(b) $m = 1$, $T = 10$

(c) $T = 20$ and the weight of the box is 12 N.

6 'Newton's cradle' consists usually of five small steel spheres, each suspended by two light strings. Modelling this by a number of particles each suspended by two strings inclined at $80°$ to the horizontal, find the tension in each string when:

(a) the weight of each steel sphere is 0.8 N

(b) the mass of each steel sphere is 0.1 kg.

7

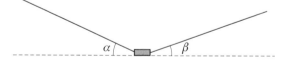

A peg-bag of mass m hanging from a washing line may be modelled by a particle supported in equilibrium by two straight strings inclined at angles α and β to the horizontal, as shown in the diagram. Find the tension in each string when:

(a) $m = 0.5$, $\alpha = \beta = 10°$

(b) $m = 0.5$, $\alpha = 10°$, $\beta = 12°$

(c) the weight of the bag is 2 N, $\alpha = 6°$, $\beta = 7°$.

Why in practice will α and β never differ by very much?

8

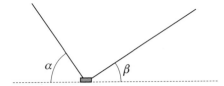

An electric light mass m kg in a workshop is held over a particular place by two strings attached to the light bulb holder. This is modelled by a particle suspended by two straight strings inclined at angles α and β above the horizontal. Find the tension in the strings when:

(a) $m = 0.5$, $\alpha = \beta = 60°$

(b) $m = 0.5$, $\alpha = 60°$, $\beta = 65°$.

4.7 Friction and the coefficient of friction

Now have a closer look at the frictional force described in section 4.6.

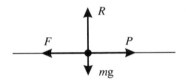

Remember that if a horizontal force **P** is applied to a particle lying on a rough horizontal table, the particle does not necessarily move. There is a frictional force **F** opposing the force **P**.

If **P** and **F** are the only horizontal forces acting on the particle then so long as the particle is stationary:

$$P = F$$

■ **The magnitude of the frictional force is just sufficient to prevent relative motion.**

The frictional force **F** for a particular surface is not constant. It increases as the applied force **P** increases until the force **F** reaches a value $\mathbf{F}_{max}$ beyond which it cannot increase. The particle is then just about to move and is said to be in a state of **limiting equilibrium**. At this point friction is said to be **limiting**.

It can be shown experimentally that F_{max} is proportional to R, that is:

$$F_{max} = \mu R$$

The constant of proportionality, which is always given as the symbol μ, is called the **coefficient of friction**. (μ is a letter of the Greek alphabet and is pronounced 'mu'.)

In general, therefore, $F \leqslant \mu R$. Clearly $F \geqslant 0$ and is only zero when the surface is completely smooth. (In real life, no surface *is* completely smooth!)

Once the particle begins to move the frictional force opposing the relative motion *remains* at the constant value μR. The value of μ depends on the nature of the two surfaces in contact.

■ To summarise:

1. **Friction acts to oppose relative motion.**
2. **Until it reaches its limiting value the magnitude of the frictional force is just sufficient to prevent relative motion.**
3. **When the limiting value is reached $F = \mu R$, where R is the normal contact force and μ is the coefficient of friction.**
4. **For all *rough* surfaces $0 < F \leqslant \mu R$.**
5. **For a *smooth* surface $F = 0$.**
6. **When the particle begins to slide *the frictional force takes its limiting value μR and acts in the direction *opposite* to the direction of relative motion.**

Example 12

A particle of weight 30 N rests on a horizontal plane. The coefficient of friction between the particle and the plane is 0.3. A horizontal force of magnitude P N is applied to the particle. Find the value of P given that the particle is about to slide.

The force diagram is:

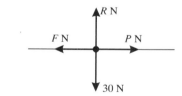

Resolving horizontally $\rightarrow$ gives:
$$P - F = 0$$
$$P = F$$

Resolving vertically $\uparrow$ gives:
$$R - 30 = 0$$
$$R = 30$$

When the particle is about to slide, friction is limiting, that is:

$$F = \mu R$$

So: $P = F = \mu R = (0.3) \times 30$

Hence: $P = 9$

Example 13

A particle of weight 30 N rests in equilibrium on a rough horizontal table. A string is attached to the particle. The string makes an angle of 30° with the horizontal and the tension in the string is 18 N. Find the magnitude of the frictional force acting on the particle.

The forces acting on the particle are the tension T along the string and the weight 30 N. As a result of the contact with the table there will be a normal reaction R N and a frictional force F N parallel to the table. The direction of F N will be such as to oppose any subsequent motion. The force diagram is:

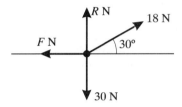

You want to find F, so resolve horizontally:

$\rightarrow$ $18 \cos 30° - F = 0$

$$F = 18 \cos 30°$$

$$= 18(\tfrac{\sqrt{3}}{2}) = 9\sqrt{3} = 15.6$$

The magnitude of the frictional force is 15.6 N.

Example 14

A particle of mass 4 kg rests in limiting equilibrium on a rough plane inclined at 30° to the horizontal. Find the coefficient of friction between the particle and the plane.
The force diagram is:

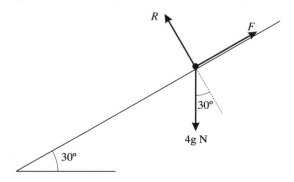

Notice that **F** acts *up* the plane, since any subsequent motion will be *down* the plane. (Things slide downhill, not uphill!)

Resolving perpendicular to the plane $\nwarrow$ gives:

$$R - 4g \cos 30° = 0$$

$$R = 4g \cos 30° = 4g \times \tfrac{\sqrt{3}}{2} = 2g\sqrt{3}$$

Resolving up the plane $\nearrow$ gives: $F - 4g \sin 30° = 0$

$$F = 4g \sin 30° = 4g \times \tfrac{1}{2} = 2g$$

As equilibrium is limiting, $F = \mu R$ and so:

$$2g = \mu \times 2g\sqrt{3}$$

$$\mu = \frac{1}{\sqrt{3}} = 0.577$$

The coefficient of friction between the particle and the plane is 0.577.

Example 15

A parcel of mass 3 kg rests in limiting equilibrium on a rough plane inclined at 30° to the horizontal. The coefficient of friction between the parcel and the plane is $\frac{1}{3}$. A horizontal force of magnitude X N is applied to the particle so that equilibrium will be broken by the particle moving upwards. Find the value of X.

The first step in the solution of this problem is to replace the parcel by a particle of mass 3 kg. The force diagram is then:

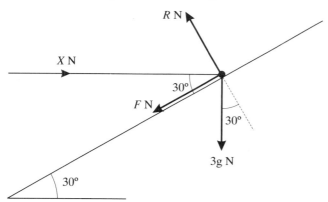

Notice that **F** is down the plane as you are told that subsequent motion will be *up* the plane.

Resolving perpendicular to the plane $\nwarrow$ gives:

$$R - X \sin 30° - 3g \cos 30° = 0$$

So:
$$R - X \times \tfrac{1}{2} - 3g \times \tfrac{\sqrt{3}}{2} = 0 \qquad (1)$$

Resolving down the plane $\diagup$ gives:

$$F + 3g \sin 30° - X \cos 30° = 0$$

So:
$$F + 3g \times \tfrac{1}{2} - X \times \tfrac{\sqrt{3}}{2} = 0 \qquad (2)$$

As the equilibrium is limiting:

$$F = \mu R = \tfrac{1}{3} R$$

Substituting this result for F in equation (2) gives:

$$\tfrac{1}{3} R + \tfrac{3g}{2} - X \tfrac{\sqrt{3}}{2} = 0$$

Substituting for R from equation (1) gives:

$$\tfrac{1}{3}\left(\tfrac{1}{2} X + 3g \tfrac{\sqrt{3}}{2}\right) + \tfrac{3g}{2} - X \tfrac{\sqrt{3}}{2} = 0$$

So:
$$\begin{aligned}
X &= \frac{g(9 + 3\sqrt{3})}{3\sqrt{3} - 1} \\
&= \frac{9.8(9 + 3\sqrt{3})}{3\sqrt{3} - 1} \\
&= 33.2
\end{aligned}$$

The value of X is 33.2.

Exercise 4E

Whenever a numerical value of g is required, take $g = 9.8\,\text{m s}^{-2}$.

1 A particle of weight mg is at rest on a rough horizontal plane.
   The coefficient of friction between the particle and the plane
   is μ. A horizontal force **P** is applied to the particle, which is
   now about to slip. Find
   (a) P, given $mg = 5\,\text{N}$, $\mu = 0.4$
   (b) μ given $mg = 4\,\text{N}$, $P = 2\,\text{N}$
   (c) mg, given $\mu = 0.3$, $P = 6\,\text{N}$.

2 A particle is in equilibrium on a rough horizontal table. A
   string is attached to the particle and held at an angle 30° to
   the horizontal. The tension in the string is T, the frictional
   force exerted by the table on the particle is F. Find
   (a) F, given $T = 25\,\text{N}$,
   (b) T, given $F = 25\,\text{N}$.

3 The particle in question 2 is about to slip on the table. Given
   that the mass of the particle is $5\,\text{kg}$, find for each part of the

question the normal reaction between the particle and the table and the coefficient of friction.

4 A particle rests in limiting equilibrium on a rough plane inclined at 20° to the horizontal. Find
(a) the frictional force exerted by the plane on the particle,
(b) the coefficient of friction between the particle and the plane.

5 A body of mass 2 kg is held in limiting equilibrium on a rough plane inclined at 20° to the horizontal by a horizontal force **X**. The co-efficient of friction between the body and the plane is 0.2. Modelling the body as a particle find X when the body is on the point of slipping
(a) up the plane,
(b) down the plane.

6 The force X in Question 5 is replaced by a force **Y** at an angle of 45° to the horizontal. Find Y when the body is on the point of slipping
(a) up the plane,
(b) down the plane.

SUMMARY OF KEY POINTS

1 Force is a vector quantity.
2 Forces can be added by using the triangle law or parallelogram rule.
3 The resultant of a system of forces is most easily found by using components.
4 A system of forces is in equilibrium if their lines of action pass through a single point and if their resultant is the zero vector.
5 The magnitude of the frictional force is just sufficient to prevent relative motion.
6 For a smooth surface there is no frictional force ($F = 0$).
7 When sliding occurs the frictional force takes its limiting value μR and opposes the relative motion

Dynamics of a particle moving in a straight line

5

What is dynamics?

Chapter 3 of this book considers the motion of a particle without any reference to how that motion was produced. Chapter 4 looks at the forces which act on a particle without reference to motion. In many cases the forces acting on a particle will cause motion. **Dynamics** is the study of the relationship between the forces acting on a moving particle and the motion of the particle.

5.1 Newton's laws of motion

Newton's first law

A particle experiences a gravitational force called its weight. If this is the only force acting on the particle it will fall to earth, as seen in chapter 3. If however the particle is resting on a surface, that surface will exert a force on it preventing it from falling. This contact force will balance the weight of the particle and no motion will take place. If the particle is initially at rest on a rough surface and is acted on by a force of sufficient magnitude to overcome any friction between the particle and the surface the particle will move along the surface. Usually the particle will be seen to gain speed – in other words it will have an acceleration. A particle moving at constant speed has no acceleration and there is no resultant force acting on it.

■ **A particle will remain at rest or will continue to move with constant velocity in a straight line unless acted on by a resultant force.**

Newton's second law

A resultant force acting on a particle will cause the particle to accelerate. The acceleration is proportional to the force producing it. The same force applied to particles of different mass will produce different accelerations. The force needed to produce a given acceleration is proportional to the mass of the body being accelerated.

Mass is measured in kilograms (kg) and the basic unit of force is the newton (N). The **newton** is defined so that a force of 1 newton produces an acceleration of $1 \, \mathrm{m \, s^{-2}}$ when applied to a particle of mass 1 kg.

A force of **F** newtons applied to a particle of mass m kg will result in an acceleration of $\mathbf{a} \, \mathrm{m \, s^{-2}}$ where:

$$\mathbf{F} = m\mathbf{a}$$

Force and acceleration are vectors because each has a magnitude and a direction. Therefore the acceleration produced is in the direction of the resultant force acting on the body.

Newton's second law can be summarised by the equation $\mathbf{F} = m\mathbf{a}$. This is often called the **equation of motion** of the particle.

■ **The force F applied to a particle is proportional to the mass m of the particle and the acceleration produced.**

$$\mathbf{F} = m\mathbf{a}$$

Newton's third law

■ **Every action has an equal and opposite reaction.**

This means that if a particle A exerts a force on a particle B then B exerts a force on A of the same magnitude but in the opposite direction. These equal and opposite forces are known as **reactions** between the particles. Newton's third law is also true when two bodies which are not both particles have contact. This law is used when studying normal contact forces – as in chapter 4. A particle at rest on the floor exerts a force on that floor. The floor also exerts a force of equal magnitude but opposite direction on the particle and the particle remains at rest.

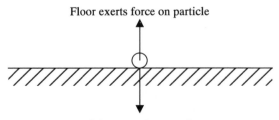

Floor exerts force on particle

Particle exerts force on floor

Example 1

Find the magnitude of the acceleration produced when a particle of mass 4 kg is acted on by a resultant force of magnitude 16 N.

The acceleration is parallel to the resultant force as shown in the diagram:

$$\longrightarrow\!\!\!\!\!\blacktriangleright \quad a\,\mathrm{m\,s^{-2}}$$

$$4\,\mathrm{kg}\ \circ\!\!\longrightarrow\quad 16\,\mathrm{N}$$

Let the acceleration be $a\,\mathrm{m\,s^{-2}}$.

Using: $F = ma$

Gives: $16 = 4 \times a$

$$a = 4$$

The magnitude of the acceleration is $4\,\mathrm{m\,s^{-2}}$.

Example 2

A particle of mass 3 kg is accelerating at $12\,\mathrm{m\,s^{-2}}$. Find the magnitude of the resultant force on the particle.

The resultant force is parallel to the acceleration.

$$\longrightarrow\!\!\!\!\!\blacktriangleright \quad 12\,\mathrm{m\,s^{-2}}$$

$$3\,\mathrm{kg}\ \circ\!\!\longrightarrow\quad F\,\mathrm{N}$$

Let the resultant force be $F\mathrm{N}$.

Using: $F = ma$

Gives: $F = 3 \times 12 = 36$

The resultant force is of magnitude 36 N.

Example 3

Find in vector form the acceleration of a particle of mass 600 g when forces of $(7\mathbf{i} + 13\mathbf{j})\,\mathrm{N}$, $(4\mathbf{i} + 4\mathbf{j})\,\mathrm{N}$ and $(-2\mathbf{i} - 5\mathbf{j})\,\mathrm{N}$ act on it. Find also the magnitude and direction of the acceleration.

You must first convert the mass of the particle to kilograms.

Mass of particle $= 600\,\mathrm{g} = 0.6\,\mathrm{kg}$, as $1\,\mathrm{kg} = 1000\,\mathrm{g}$.

Resultant force $= (7\mathbf{i} + 13\mathbf{j}) + (4\mathbf{i} + 4\mathbf{j}) + (-2\mathbf{i} - 5\mathbf{j})$

$$= 9\mathbf{i} + 12\mathbf{j}$$

Using: $\mathbf{F} = m\mathbf{a}$

Gives: $9\mathbf{i} + 12\mathbf{j} = 0.6\,\mathbf{a}$

$$\mathbf{a} = \frac{9\mathbf{i} + 12\mathbf{j}}{0.6}$$

$$= 15\mathbf{i} + 20\mathbf{j}$$

The acceleration is $(15\mathbf{i} + 20\mathbf{j})\,\mathrm{m\,s^{-2}}$.

The magnitude of **a** is:

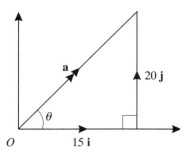

$$|\mathbf{a}| = \sqrt{(15^2 + 20^2)} = 25$$

Also: $\tan\theta = \dfrac{20}{15}$

$$\theta = 53.1°$$

The acceleration has magnitude $25\,\mathrm{m\,s^{-2}}$ and is at an angle of $53.1°$ to the vector **i**.

Example 4

A car travels a distance of 32 m along a straight road while uniformly accelerating from rest to $16\,\mathrm{m\,s^{-1}}$. By modelling the car as a particle, find the acceleration of the car. Given that the mass of the car is 640 kg, find the magnitude of the accelerating force.

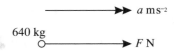

First apply the uniform acceleration formulae to find the acceleration of the car.

Known quantities are:

$$s = 32 \text{ m}$$
$$u = 0\,\mathrm{m\,s^{-1}}$$
$$v = 16\,\mathrm{m\,s^{-1}}$$

Using: $v^2 = u^2 + 2as$

Substituting for v, u and s gives:

$$16^2 = 0^2 + 2 \times a \times 32$$

$$a = \frac{16^2}{2 \times 32} = 4$$

The acceleration of the car is $4\,\mathrm{m\,s^{-2}}$.

To find the magnitude of the accelerating force you must use the equation of motion for the car.

$$F = ma$$
$$F = 640 \times 4 = 2560$$

The accelerating force has magnitude $2560\,\text{N}$.

Example 5

A car of mass $600\,\text{kg}$ is travelling along a straight horizontal road with an acceleration of $2\,\text{m}\,\text{s}^{-2}$. The engine is exerting a forward force of magnitude $1500\,\text{N}$. By modelling the car as a particle, find the magnitude of the resistance it is experiencing.

Let the resistance be R newtons, as shown in the diagram.
The resultant force on the car in the direction of motion is $(1500 - R)\,\text{N}$.

Using: $\qquad\qquad\qquad F = ma$

Substituting for F gives:

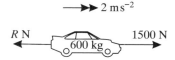

$$1500 - R = 600 \times 2$$
$$R = 1500 - 1200 = 300$$

The magnitude of the resistance is $300\,\text{N}$.

Exercise 5A

1 Find the acceleration produced when a particle of mass $2\,\text{kg}$ is acted on by a resultant force of $12\,\text{N}$.

2 Find the resultant force which will produce an acceleration of $5\,\text{m}\,\text{s}^{-2}$ for a particle of mass $6\,\text{kg}$.

3 A particle of mass $5\,\text{kg}$ has an acceleration of $4\,\text{m}\,\text{s}^{-1}$. Find the magnitude of the resultant force acting on the particle.

4 A particle of mass $4\,\text{kg}$ experiences a resultant force of $20\,\text{N}$. Find the acceleration produced.

5 Find the resultant force which will produce an acceleration of $2\,\text{m}\,\text{s}^{-2}$ for a particle of mass $5\,\text{kg}$.

6 A particle has an acceleration of $3\,\text{m}\,\text{s}^{-2}$ when acted on by a resultant force of $12\,\text{N}$. Find the mass of the particle.

7 A particle of mass $5\,\text{kg}$ is acted on by a force of $(4\mathbf{i} + 6\mathbf{j})\,\text{N}$. Find in vector form the resulting acceleration.

8 A particle of mass $2\,\text{kg}$ is accelerating at $(4\mathbf{i} - 2\mathbf{j})\,\text{m}\,\text{s}^{-2}$. Find in vector form the resultant force acting on the particle.

9 A particle of mass 2 kg is acted on by forces of $(5\mathbf{i} - 2\mathbf{j})$ N and $(3\mathbf{i} + 4\mathbf{j})$ N. Find in vector form the acceleration produced.

10 A packing case of mass 60 kg is dragged across a rough horizontal floor by a horizontal force of 50 N. The acceleration of the case is $0.4\,\mathrm{m\,s^{-2}}$. Model the packing case as a particle and hence find the magnitude of the friction force.

11 A car travels a distance of 30 m along a straight horizontal road while uniformly accelerating from rest to $15\,\mathrm{m\,s^{-1}}$. Find the acceleration of the car. Given that the car has a mass of 560 kg find the magnitude of the accelerating force.

12 A toy dog of mass 0.5 kg is pulled across a smooth horizontal floor by a horizontal string attached to the dog. The tension in the string is 0.75 N. By modelling the dog as a particle calculate the acceleration produced. Given that the dog starts from rest, determine how far it will move in 2 seconds.

13 A car of mass 550 kg is travelling along a straight horizontal road with an acceleration of $2.2\,\mathrm{m\,s^{-2}}$. The force exerted by the engine is 1600 N. Find the magnitude of the resistance to motion.

14 Forces of $(9\mathbf{i} + 3\mathbf{j})$ N, $(7\mathbf{i} + 3\mathbf{j})$ N and $(a\mathbf{i} + b\mathbf{j})$ N, where a and b are constants, act on a particle of mass 2 kg. Given that the acceleration produced is $(10\mathbf{i} + 2\mathbf{j})\,\mathrm{m\,s^{-2}}$ find the values of a and b.

15 A lorry of mass 2.5 tonnes experiences resistances totalling 800 N when travelling along a level road. The engine is producing a driving force of 2400 N. By modelling the lorry as a particle find the acceleration of the lorry.

16 A car of mass 600 kg is brought to rest in 6 seconds from a speed of $20\,\mathrm{m\,s^{-1}}$. Neglecting resistances find the braking force required to achieve this.

17 A car of mass 600kg experences a resistive force of 500 N while being brought to rest in 6 seconds from a speed of $20\,\mathrm{m\,s^{-1}}$. Calculate the braking force required to achieve this.

18 Given that the resistances total 400 N find the magnitude of the constant force needed to accelerate a car of mass 800 kg from rest to $20\,\mathrm{m\,s^{-1}}$ in 15 s.

19 Find the magnitude of the resultant force required to give a particle of mass 3 kg an acceleration of $(2\mathbf{i} - 3\mathbf{j})\,\mathrm{m\,s^{-2}}$.

20 A stone slides in a straight line across a frozen pond. Given that the initial speed of the stone is $5\,\mathrm{m\,s^{-1}}$ and that it slides 20 m before coming to rest, calculate the coefficient of friction between the stone and the surface of the frozen pond.

Vertical motion

When a particle is moving vertically, one of the forces acting in the direction of motion is the weight of the particle. A particle falling freely under gravity has an acceleration of $\mathrm{g} = 9.8\,\mathrm{m\,s^{-2}}$. Because the only force acting on the particle is its weight, the equation of motion is:

■
$$\mathbf{F} = \mathbf{mass} \times \mathbf{g}$$
or:
$$\mathbf{weight} = m\mathbf{g}$$

Example 6

A particle of mass 2 kg is attached to the lower end of a string hanging vertically. The particle is lowered and moves with an acceleration of $0.2\,\mathrm{m\,s^{-2}}$. Find the tension in the string.

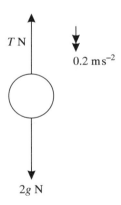

Mass of particle = 2 kg.

Therefore weight of particle = 2g N.

Let the tension in the string be T N as shown in the diagram.

Because the particle is moving downwards, the resultant force acting on the particle is downwards.

So the resultant downwards force $= (2\mathrm{g} - T)$ N

Using: $F = ma$

Substituting for F gives: $2\mathrm{g} - T = 2 \times 0.2$
$$T = 2 \times 9.8 - 2 \times 0.2 = 19.2$$

The tension in the string is 19.2 N.

Example 7

A stone of mass 0.5 kg is released from rest on the surface of the water in a well. It takes 2 seconds to reach the bottom of the well.

Given that the water exerts a constant resistance of 2 N, find the depth of the well.

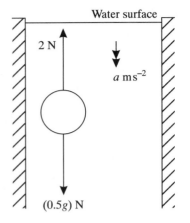

Using: $$F = ma$$
Gives: $$0.5\,g - 2 = 0.5a$$
$$a = \frac{0.5 \times 9.8 - 2}{0.5} = 5.8$$

The acceleration of the particle is $5.8\,\mathrm{m\,s^{-2}}$.

The forces acting on the stone are constant throughout the motion. This means that the acceleration is also constant and the uniform acceleration equations can be applied.

Known quantities are:

$$u = 0\,\mathrm{m\,s^{-1}}$$
$$a = 5.8\,\mathrm{m\,s^{-2}}$$
$$t = 2\,\mathrm{s}$$

Using: $$s = ut + \tfrac{1}{2}at^2$$

Substituting for u, a and b gives:

$$s = 0 + \tfrac{1}{2} \times 5.8 \times 2^2$$
$$= 11.6$$

The well is 11.6 m deep.

Example 8

A parcel of mass 5 kg is released from rest on a rough ramp of inclination $\theta = \arcsin \tfrac{3}{5}$ (that is $\sin\theta = \tfrac{3}{5}$) and slides down the ramp. After 3 seconds the parcel has a speed of $4.9\,\mathrm{m\,s^{-1}}$. Treating the parcel as a particle, find the coefficient of friction between the parcel and the ramp (see chapter 4 for work on the coefficient of friction).

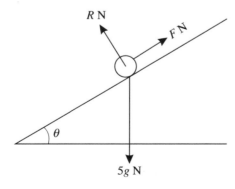

R N is the reaction of the ramp on the parcel.

The component of the weight perpendicular to the ramp is $5g \cos \theta$ N.

There is no motion perpendicular to the ramp so resolving perpendicular to the ramp gives:

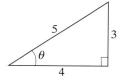

$$R - 5g \cos \theta = 0$$
$$R = 5g \cos \theta = 5g \times \tfrac{4}{5} = 4g$$

The particle has an acceleration parallel to the ramp.

The friction force acts up the ramp to oppose the motion.

The component of weight down the ramp is $5g \sin \theta$ N. Using the equation of motion down the ramp gives:

$$5g \sin \theta - F = 5a \qquad\qquad (1)$$

Use the uniform acceleration equations to find a.

Known quantities are:

$$t = 3\,\text{s}$$
$$u = 0\,\text{m s}^{-1}$$
$$v = 4.9\,\text{m s}^{-1}$$

Using:
$$v = u + at$$

Substituting for t, u and v gives:

$$4.9 = 0 + 3\,a$$
$$a = \frac{4.9}{3}$$

Substituting this value for a in equation (1) and using $\sin \theta = \tfrac{3}{5}$ gives:

$$F = 5g \times \tfrac{3}{5} - 5 \times \frac{4.9}{3}$$
$$F = 3 \times 9.8 - \frac{5 \times 4.9}{3}$$
$$F = 21.23 \text{ and from above } R = 4g$$

For sliding $F = \mu R$

$$\mu = \frac{F}{R}$$

Substituting $R = 4g$ and $F = 21.23$ gives:

$$\mu = \frac{21.23}{4 \times 9.8} = 0.5415$$

The coefficient of friction is 0.542.

Example 9

A lift is accelerating upwards at $1.5 \, \text{m s}^{-2}$. A child of mass 30 kg is standing in the lift. Treating the child as a particle find the force between the child and the floor of the lift.

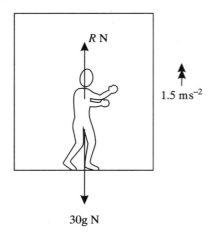

The child exerts a force on the floor of the lift.

By Newton's third law, the floor exerts an equal and opposite force on the child. Let this force be R N as shown in the diagram.

The resultant upwards force on the child is $(R - 30 \, g) \, \text{N}$.

Using: $\qquad\qquad\qquad F = ma$

Gives: $\qquad\qquad R - 30 \, g = 30 \times 1.5$

$\qquad\qquad\qquad\quad R = 30 \times 1.5 + 30 \times 9.8$

$\qquad\qquad\qquad\quad R = 339$

The force between the child and the floor is 339 N.

Exercise 5B

Whenever a numerical value of g is required take $g = 9.8 \, \text{m s}^{-2}$.

1 A particle of mass 0.3 kg is attached to the lower end of a vertical string. If the particle has a downwards acceleration of magnitude $0.9 \, \text{m s}^{-2}$ find the tension in the string.

2 A stone of mass 0.06 kg is falling through a liquid with an acceleration of $3.6 \, \text{m s}^{-2}$. Find the resistive force acting on the stone.

3 A boy of mass 50 kg travels up in a lift. The acceleration of the lift is $0.4 \, \text{m s}^{-2}$. Find the force exerted on the boy by the

floor of the lift. What force does the boy exert on the floor of the lift?

4 A cable raising a load with an acceleration of $1.5 \, \text{m s}^{-2}$ has a tension of 15 kN. Find the mass of the load.

5 A boy is tobogganing down a smooth slope inclined at 20° to the horizontal. Find his acceleration, assuming resistances can be neglected.

6 A boy is tobogganing down a slope inclined at 25° to the horizontal. The resistances to his motion amount to 15 N. By modelling the boy and his toboggan as a single particle, find the mass of the boy and his toboggan when his acceleration is $3.9 \, \text{m s}^{-2}$.

7 A particle of mass 9 kg is sliding down a smooth inclined plane with an acceleration of $4.9 \, \text{m s}^{-2}$. Find the angle of inclination of the plane.

8 A particle of mass 6 kg is sliding down a smooth plane inclined at 45° to the horizontal. Find the acceleration of the particle.

9 A particle of mass 2 kg is sliding down a plane inclined at 24° to the horizontal against a resistive force of 3 N. Find the acceleration of the particle.

10 A parcel of mass 3 kg is sliding down a rough slope of inclination arcsin 0.4. The coefficient of friction between the parcel and the slope is 0.35. By modelling the parcel as a particle find its acceleration.

11 A parcel of mass 60 kg is released from rest on a smooth plane inclined at arcsin $\frac{3}{5}$ to the horizontal. Find the velocity of the particle when it has travelled 5 m down the plane.

12 A particle of mass 2 kg slides down a rough plane inclined at 20° to the horizontal. Given that the acceleration of the particle is $1.5 \, \text{m s}^{-2}$ find the coefficient of friction between the particle and the plane.

13 A particle of mass 4 kg is being pulled up a rough plane inclined at 25° to the horizontal by a force of magnitude 30 N acting along a line of greatest slope of the plane. Given that the particle is accelerating at $2 \, \text{m s}^{-2}$ find the coefficient of friction between the particle and the plane.

14 A package of mass 10 kg is released from rest on a rough slope inclined at 25° to the horizontal. After 2 seconds the package has moved 4 m down the slope. Find the coefficient of friction between the package and the slope.

15 A horizontal force of 2 N is just sufficient to prevent a block of mass 1 kg from sliding down a rough plane inclined at arc sin $\frac{7}{25}$ to the horizontal. Find the coefficient of friction between the block and the plane and the acceleration with which the block will move when the force is removed.

16 A particle rests in **limiting equilibrium** (that is the particle is on the point of moving and friction has its maximum value) on a plane inclined at 30° to the horizontal. Determine the acceleration with which the particle will slide down the plane when the angle of inclination is increased to 40°.

17 A particle of mass 2 kg rests in limiting equilibrium on a plane inclined at 25° to the horizontal. The angle of inclination is decreased to 20° and a force of magnitude 20 N is applied up a line of greatest slope. Find the particle's acceleration. When the particle has been moving for 2 seconds the force is removed. Determine the further distance the particle will move up the plane.

18 A block of mass 1.6 kg is placed on a rough plane inclined at 45° to the horizontal. The coefficient of friction between the block and the plane is $\frac{1}{4}$. Model the block as a particle and hence find the acceleration of the block down the plane. Find the velocity of the block after 2 seconds, assuming that it starts from rest.

5.2 The motion of two connected particles

When two moving particles are connected by a string which is light and inextensible there will be a tension in the string. By Newton's third law, the forces acting on the particles will have the same magnitude but will act in opposite directions as shown in the diagram.

It is important on diagrams to show these separate tensions clearly.

It will be assumed that all strings are light and inextensible.

Example 10

Two particles A and B of masses $3\,kg$ and $4\,kg$ respectively connected by a light inextensible string are at rest on a smooth horizontal surface. A force of magnitude $7\,N$ is applied to particle B in the direction AB. Find the acceleration produced and the tension in the string.

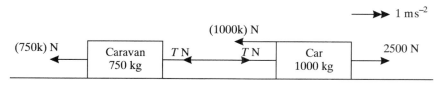

The motion of the particles A and B need to be considered separately. Suppose each has an acceleration of $a\,m\,s^{-2}$. Let the tension in the string be $T\,N$ as shown in the diagram.

Using:

$$F = ma$$

Equation of motion for B is:

$$7 - T = 4a \qquad (1)$$

Equation of motion for A is:

$$T = 3a \qquad (2)$$

Adding the simultaneous equations (1) and (2) gives:

$$7 = 4a + 3a$$
$$a = 1$$

The acceleration of either particle is $1\,m\,s^{-2}$.

Substituting $a = 1$ in equation (2) gives

$$T = 3 \times 1 = 3$$

The tension in the string is $3\,N$.

Example 11

A car of mass $1000\,kg$ tows a caravan of mass $750\,kg$ along a horizontal road. The engine of the car exerts a forward force of $2.5\,kN$. The resistances to the motion of the car and caravan are each $k \times$ their mass where k is constant. Given that the car accelerates at $1\,m\,s^{-2}$ find the tension in the tow-bar.

The diagram shows the information given.

Using $F = ma$ the equation of motion for the car is:

$$2500 - 1000\,\text{k} - T = 1000 \times 1$$
$$1000\,\text{k} + T = 2500 - 1000$$
$$1000\,\text{k} + T = 1500 \qquad (1)$$

Equation of motion for the caravan is:

$$T - 750\,\text{k} = 750 \times 1$$

$$T - 750\,\text{k} = 750 \qquad (2)$$

Multiplying equation (1) by 3 gives:

$$3000\,\text{k} + 3T = 4500 \qquad (3)$$

Multiplying equation (2) by 4 gives:

$$4T - 3000\,\text{k} = 3000 \qquad (4)$$

Adding equations (3) and (4) gives:

$$7T = 7500$$
$$T = \frac{7500}{7} = 1071$$

The tension in the tow-bar is 1070 N.

Exercise 5C

Whenever a numerical value of g is required take $g = 9.8\,\text{m s}^{-2}$.

1 A car of mass 1000 kg is towing a caravan of mass 600 kg along a horizontal road. Given that the driving force produced by the engine is 400 N and that there is no resistance to motion find the tension in the tow-bar and the acceleration of the car.

2 A car of mass 850 kg is towing a caravan of mass 550 kg along a level road. The engine of the car exerts a forward force of 2 kN. The resistances to the motion of the car and caravan are proportional to their masses – in other words the resistances are k × their masses, where k is a constant. Given that the car accelerates at $0.5\,\text{m s}^{-2}$ find the tension in the tow-bar.

3 The diagram shows a particle A of mass 0.5 kg suspended by a vertical string. A particle B of mass 0.4 kg is suspended from A by means of another string. A force of 10 N is applied

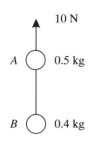

to the upper string and the particles move upwards. Find the tension in the lower string and the acceleration of the system.

4 A car of mass 900 kg tows a caravan of mass 500 kg against resistances totalling 700 N. The resistances on the car and caravan are proportional to their masses – in other words the resistances are k × their masses where k is a constant. The car is accelerating at $0.8\,\mathrm{m\,s}^{-2}$ along a horizontal road. By modelling the car and caravan as a pair of connected particles, find the driving force exerted by the engine and the tension in the tow-bar.

5 A car of mass 1000 kg exerts a driving force of 2.2 kN when pulling a caravan of mass 500 kg along a horizontal road. The car and caravan increase speed from rest to $4\,\mathrm{m\,s}^{-1}$ while travelling 16 m. Given that the resistances on the car and caravan are proportional to their masses, find these resistances and the tension in the tow-bar.

6 The diagram shows a block A of mass 100 kg suspended by a vertical cable. A block B of mass 150 kg is suspended from A by means of a second vertical cable. The blocks are raised 10 m in 10 seconds, starting from rest. Find the tension in each cable.

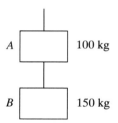

Problems involving pulleys

The figure shows two particles joined by a light inextensible string which passes over a fixed pulley. If the two particles are of different masses then the heavier particle will move vertically downwards and the lighter particle will move vertically upwards. However, the motions of the two particles are not independent because they are connected by the string. This is the motion of connected particles again although this time not along a single straight line.

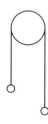

Example 12

Particles of mass 5 kg and 2 kg are attached to the ends of a light inextensible string which passes over a smooth fixed pulley. The system is released from rest. Find the acceleration of the system and the distance moved by the 5 kg mass in the first 3 seconds of the motion. (Assume that neither particle reaches the pulley.)

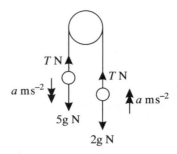

Since the particles are released from rest the heavier particle will move downwards. Let the particle's acceleration be $a\,\text{m}\,\text{s}^{-2}$. As the particles are joined by a light inextensible string, the lighter mass will have an acceleration $a\,\text{m}\,\text{s}^{-2}$ upwards.

Let the tension in the string be $T\,\text{N}$. It will be the same throughout the string as the pulley is smooth.

So for the motion of the particle of mass $5\,\text{kg}$ use:

$$F = ma$$

Giving: $$5g - T = 5a \qquad (1)$$

And for the motion of the particle of mass $2\,\text{kg}$ use:

$$F = ma$$

Giving: $$T - 2g = 2a \qquad (2)$$

Adding equations (1) and (2) gives:

$$5g - 2g = 5a + 2a$$
$$7a = 3g$$
$$a = \frac{3g}{7} = 3 \times \frac{9.8}{7} = 4.2$$

The acceleration is $4.2\,\text{m}\,\text{s}^{-2}$.

To find the distance moved by the $5\,\text{kg}$ particle in 3 seconds use the constant acceleration equations.

Known quantities are:

$$u = 0\,\text{m}\,\text{s}^{-1}$$
$$a = 4.2\,\text{m}\,\text{s}^{-2}$$
$$t = 3\,\text{s}$$

Using: $$s = ut + \tfrac{1}{2}at^2$$
Gives: $$s = 0 \times 3 + \tfrac{1}{2} \times 4.2 \times 3^2$$
$$s = 18.9$$

The $5\,\text{kg}$ mass moves $18.9\,\text{m}$ in the first 3 seconds.

Example 13

Two particles P and Q of masses $6\,\mathrm{kg}$ and $3\,\mathrm{kg}$ respectively are connected by a light inextensible string. Particle P rests on a rough horizontal table. The string passes over a smooth pulley fixed at the edge of the table and Q hangs vertically. The coefficient of friction between P and the table is $\frac{1}{3}$. The system is released from rest. Find in terms of g (a) the acceleration of Q (b) the tension in the string and (c) the force exerted on the pulley.

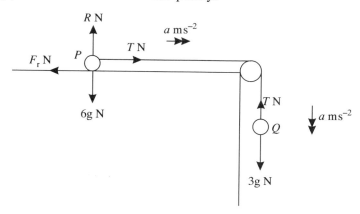

The tension in each part of the string is the same since the pulley is smooth. Let this tension be $T\,\mathrm{N}$.

Let the acceleration of the particles be $a\,\mathrm{m\,s^{-2}}$ in the directions shown in the diagram.

(a) Particle P does not move vertically. So resolving vertically gives:
$R - 6\,\mathrm{g} = 0.$

As P is moving along the table, friction is limiting.

So: $F_r = \mu R = \frac{1}{3} \times 6\,\mathrm{g} = 2\,\mathrm{g}$

Using: $F = ma$ along the table

Gives: $T - F_r = 6a$
$$T - 2\,\mathrm{g} = 6a \qquad (1)$$

Resolving vertically for Q using: $F = ma$

Gives: $3\mathrm{g} - T = 3a \qquad (2)$

Adding equations (1) and (2) gives:

$$3\,\mathrm{g} - 2\,\mathrm{g} = 6a + 3a$$
$$9a = \mathrm{g}$$
$$a = \frac{\mathrm{g}}{9}$$

The acceleration of Q is $\dfrac{\mathrm{g}}{9}\,\mathrm{m\,s^{-2}}$ downwards.

(b) Substituting $a = \dfrac{g}{9}$ in equation (1) to find the tension in the string gives:

$$T - 2g = 6 \times \frac{g}{9}$$

$$T = 2g + \frac{2g}{3}$$

$$T = \frac{8g}{3}$$

The tension in the string is $\dfrac{8g}{3}$ N.

(c) To find the force exerted on the pulley consider the horizontal and vertical parts of the string separately. The tension in each part of the string is the same.

By Newton's third law, there will be equal but opposite tensions at the pulley ends of the strings:

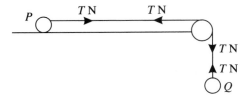

The forces on the pulley are:

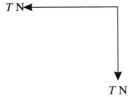

where $T = \dfrac{8g}{3}$ from (b) above.

To calculate the resultant apply the triangle law of addition (see chapter 2).

This gives:

Magnitude of resultant $= \sqrt{\left[\left(\dfrac{8g}{3}\right)^2 + \left(\dfrac{8g}{3}\right)^2 \right]} = \dfrac{8g}{3}\sqrt{2}$

$\theta = 45°$ as the triangle is isosceles.

The resultant force on the pulley is of magnitude $\dfrac{8g}{3}\sqrt{2}$ N in a direction 45° below the horizontal.

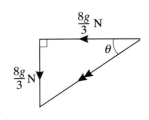

Example 14

A particle P of mass 5 kg lies on a smooth inclined plane of angle $\theta = \arcsin \frac{3}{5}$ (that is $\sin \theta = \frac{3}{5}$). Particle P is connected to a particle Q of mass 4 kg by a light inextensible string which lies along a line of greatest slope of the plane and passes over a smooth light pulley. The system is held at rest with Q hanging vertically 2 m above a horizontal plane. The system is now released from rest. Assuming P does not reach the pulley, find, to 3 significant figures:

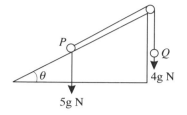

(a) the initial acceleration of Q
(b) how long it takes for Q to hit the horizontal plane
(c) the total distance that P moves up the plane.

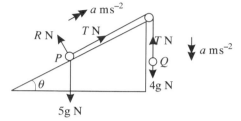

Using $\sin \theta = \frac{3}{5}$, from the (3, 4, 5) triangle, $\cos \theta = \frac{4}{5}$.

(a) As the pulley is smooth, the tension is the same throughout the string. Let this tension be T N.

Let the acceleration of Q be a m s^{-2} downwards.

The acceleration of P is a m s^{-2} up the plane.

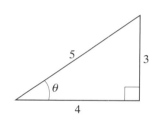

The plane is smooth so there is no friction force.

The weight of P has component $5 g \sin \theta$ N parallel to the plane.

Using:
$$F = ma$$

For P, parallel to the plane the equation of motion is:

$$T - 5 g \sin \theta = 5a$$

$$T - 5 g \times \frac{3}{5} = 5a$$

$$T - 3 g = 5a \qquad (1)$$

For Q vertically the equation of motion is:

$$4 g - T = 4a \qquad (2)$$

Adding equations (1) and (2) gives:

$$4 g - 3 g = 5a + 4a$$
$$a = \frac{g}{9} = \frac{9.8}{9} = 1.088$$

The initial acceleration is 1.09 m s^{-2}.

(b) Using: $s = ut + \frac{1}{2}at^2$

Known quantities are $s = 2\,\mathrm{m}$, $u = 0$, and $a = \dfrac{9.8}{9}\,\mathrm{m\,s^{-2}}$

Gives:
$$2 = 0 \times t + \frac{1}{2} \times \frac{9.8}{9}t^2$$
$$t^2 = \frac{2 \times 2 \times 9}{9.8}$$
$$t = 1.916$$

Q takes $1.92\,\mathrm{s}$ to reach the horizontal plane.

(c) To calculate the total distance P travels up the plane before coming to rest separate P's motion up the plane into two parts, AB and BC.

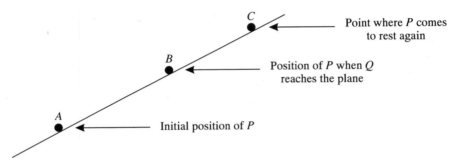

AB is the distance moved from the start of motion to the moment when Q reaches the plane.

BC is the distance moved from the moment Q reaches the plane until P comes to rest.

From A to B motion is as connected particles. Because Q descends $2\,\mathrm{m}$ before hitting the plane, P must travel the same distance so $AB = 2\,\mathrm{m}$.

Finding the speed of P when it reaches B will give the initial speed for the motion from B to C.

Using: $v^2 = u^2 + 2as$

Known quantities are $u = 0$, $a = 1.088\,\mathrm{m\,s^{-2}}$ and $s = 2\,\mathrm{m}$

Gives: $v^2 = 0 + 2 \times 1.088 \times 2$

$v = 2.086$

On reaching B, P has speed $2.086\,\mathrm{m\,s^{-1}}$ up the plane and has travelled $2\,\mathrm{m}$.

For the motion from B to C, a new diagram is needed because when Q reaches the plane the string becomes slack. The only forces on P are its weight and the reaction between P and the plane.

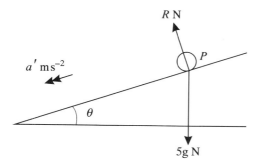

As the forces on P have changed, so will the acceleration.

Let the new acceleration be $a'\,\text{m s}^{-2}$.

Using: $$F = ma$$

The equation of motion parallel to the plane is:

$$5\,g\sin\theta = 5a'$$

$$a' = \frac{3}{5}\,g = 5.88 \text{ down the plane}$$

The motion from B to C is up the plane so the particle has an acceleration of -5.88 m s^{-2}.

Using: $\qquad v^2 = u^2 + 2as \quad$ with $\quad a = -5.88\,\text{m s}^{-2}$,

$$u = 2.086\,\text{m s}^{-1} \qquad v = 0\,\text{m s}^{-1}$$

Gives: $\qquad\qquad 0 = 2.086^2 - 2 \times 5.88s$

$$s = \frac{2.086^2}{2 \times 5.88} = 0.370$$

So the distance $BC = 0.37\,\text{m}$.

Therefore the total distance moved by P is $AB + BC = 2.37\,\text{m}$.

Exercise 5D

Whenever a numerical value of g is required take $g = 9.8\,\text{m s}^{-2}$.

1 Particles of mass 4 kg and 2 kg are attached to the ends of a light inextensible string which passes over a smooth fixed pulley. The system is released from rest. Find the acceleration of the system and the tension in the string.

2 Particles of mass 4 kg and 7 kg are attached to the ends of a light inextensible string which passes over a smooth fixed pulley. Find the acceleration of the system and the force exerted on the pulley.

3 Particles of mass $2M$ kg and M kg are connected by a light inextensible string which passes over a smooth fixed pulley. The system is released from rest with the hanging parts of the string vertical. Find the acceleration of the system in terms of g.

4 Two particles A and B are connected by a light inextensible string which passes over a smooth fixed pulley. A has mass 4 kg and descends 2.8 m in the first 2 seconds after the system is released from rest, the hanging parts of the string being vertical. Find the mass of B and the tension in the string.

5 Two particles P and Q of masses 3 kg and 6 kg respectively are attached to the ends of a light inextensible string. The string passes over a smooth fixed pulley. The system is released from rest with both masses a distance of 2 m above a horizontal floor. Find how long it takes for particle Q to hit the floor. Assuming particle P does not reach the pulley find its greatest height above the floor.

6 A rope is slung over a smooth beam. Two children are hanging on to the ends of the rope. The heavier child has a mass of 56 kg and descends with an acceleration of $2\,\mathrm{m\,s^{-2}}$. By modelling the children as a pair of connected particles find the tension in the rope and the mass of the lighter child.

7 Two particles P and Q of masses 7 kg and 2 kg respectively are connected by a light inextensible string. Particle P rests on a smooth table. The string passes over a smooth pulley fixed at the edge of the table and Q is hanging freely. The system is released from rest. Find the acceleration of P.

8 Given that the table for the system described in Question 7 is rough instead of smooth and that the coefficient of friction between P and the table is $\frac{1}{7}$, find the acceleration of the system and the tension in the string.

9 Two particles P and Q of masses 0.8 kg and 0.6 kg respectively are connected by a light inextensible string.

Particle P rests on a rough horizontal table 0.7 m from a smooth fixed pulley at the edge of the table. The string passes over the pulley and Q hangs freely, 0.5 m above the floor. The system is released from rest. The coefficient of friction between P and the table is 0.25. Find (a) the acceleration of the system (b) the tension in the string (c) the velocity with which Q hits the floor (d) the distance P moves after Q has hit the floor.

10 Two particles P and Q of masses 1 kg and 2 kg respectively are hanging vertically from the ends of a light inextensible string which passes over a smooth fixed pulley. The system is released from rest with both particles a distance of 1.5 m above a floor. When the masses have been moving for 0.5 s the string breaks. Find the further time that elapses before P hits the floor.

11 Two particles A and B of masses 0.5 kg and 0.7 kg respectively are connected by a light inextensible string. Particle A rests on a smooth horizontal table 1 m from a fixed smooth pulley at the edge of the table. The string passes over the pulley and B hangs freely 0.75 m above the floor. The system is released from rest with the string taut. After 0.5 s the string breaks. Find (a) the distance the particles have moved when the string breaks (b) the velocity of the particles when the string breaks (c) the further time that elapses before A reaches the floor, given that the table is 0.9 m high.

12 A particle A of mass 2.5 kg which is at rest on a smooth inclined plane of angle 25° is connected to a particle B of mass 1.5 kg by a light inextensible string which lies along a line of greatest slope of the plane and passes over a fixed smooth pulley at the top of the plane. B hangs freely and the system is released from rest with the string taut. Find the acceleration of the system and the tension in the string. If the angle of inclination of the plane is increased to 45° find the new acceleration.

13 A particle A of mass 0.5 kg rests on a smooth inclined plane of angle arcsin $\frac{3}{5}$. A light inextensible string is attached to this particle and passes over a smooth pulley at the top of the

plane. A particle B of mass m kg hangs freely from the other end of the string. The system is released from rest with the string taut. Particle A descends 1 m in 1 s. Find the value of m and the tension in the string.

14 A particle P of mass 5 kg rests on a rough inclined plane of angle arcsin $\frac{3}{5}$. The coefficient of friction between P and the plane is $\frac{1}{8}$. P is connected to a particle Q of mass 4 kg by a light inextensible string which lies along a line of greatest slope of the plane and passes over a smooth fixed pulley at the top of the plane. Q is hanging freely 2 m above a horizontal plane. The system is released from rest with the string taut. Assuming P does not reach the pulley find (a) the acceleration of the system (b) the time that elapses before Q hits the horizontal plane (c) the total distance P moves up the plane.

15 A particle A of mass 5 kg rests on a rough plane inclined at an angle of 30° to the horizontal. A string attached to A lies along a line of greatest slope of the plane and passes over a smooth pulley at the top of the plane. A particle B of mass 6 kg hangs vertically from the string 1 m above a horizontal plane. The system is released from rest with the string taut. If B takes 2 s to reach the horizontal plane, find the coefficient of friction between A and the inclined plane. Find also the total distance that A moves up the plane, assuming A does not reach the pulley.

5.3 Work, energy and power

Work, energy and power are words most people are familiar with and use frequently without fully understanding their meaning. All three quantities result from the application of forces to objects and so have mechanical definitions as shall be seen in this section.

Work

If the point of application of a force F newtons moves through a distance s metres in the direction of the force then the **work done by the force** is given by:

- $$Work = F \times s$$

For a force in Newtons and a distance in metres, the work done is measured in joules, abbreviated J.

Example 15

A packing case is pushed 5 m across a floor by a horizontal force of magnitude 30 N. Find the work done by the force.

Using: $\qquad Work = F \times s$

Gives: $\qquad Work = 30 \times 5 = 150$

The work done by the force is 150 J.

Example 16

The figure shows a box which is pulled at a constant speed across a horizontal surface by a horizontal rope. When the box has moved a distance of 9 m the work done is 54 J. Find the constant resistance to the motion.

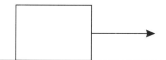

Because the speed is constant there is no acceleration.

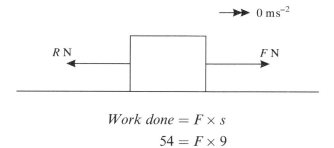

$$Work\ done = F \times s$$
$$54 = F \times 9$$
$$F = 6$$

Using the equation of motion: $F - R = 0$

Gives: $\qquad\qquad\qquad R = 6$

The resistance to motion is 6 N.

Work done against gravity

To raise a particle of mass m kg vertically at a constant speed you need to apply a force of mg newtons vertically upwards. If the particle is raised a distance of h metres the work done against gravity is mgh joules. Work is done against gravity when the particle is moving either vertically or at an angle to the horizontal, for example on an inclined plane, but not when the particle is moving along a horizontal plane.

Example 17

An angler catches a fish of mass 1.5 kg. Find the work he does against gravity in raising the fish 4 m from sea level to the pier, assuming he winds in his line at a constant speed.

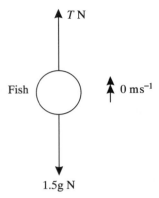

Modelling the fish as a particle moving vertically:

Since the speed is constant there is no acceleration.

Using the equation of motion vertically gives:

$$T - 1.5\,g = 0$$
$$T = 1.5\,g$$

Work done by angler against gravity $= F \times s$ and gives:

$$= 1.5 \times 9.8 \times 4$$
$$= 58.8$$

The work done by the angler is 58.8 J.

Work done against friction

When a particle is pulled up a rough inclined plane at a constant speed work must be done **against** the frictional force acting on the particle as well as against gravity (see example 18).

Example 18

A particle of mass 5 kg is pulled at constant speed a distance of 24 m up a rough plane which is inclined at 40° to the horizontal. The coefficient of friction between the particle and the surface is 0.25. Assuming the particle moves up a line of greatest slope, find:

(a) the work done against friction
(b) the work done against gravity.

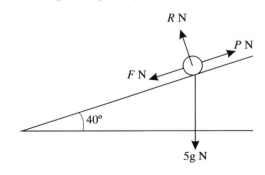

Let R be the reaction of the particle on the slope. Let F be the friction force.

(a) Because there is no motion perpendicular to the plane:

$$R - 5\,\text{g}\cos 40 = 0$$

Using $F = \mu R$ with $\mu = 0.25$ and $R = 5\,\text{g}\cos 40$ gives:

$$F = 0.25 \times 5\,\text{g}\cos 40$$

The particle moves a distance of 24 m parallel to the direction of the friction force.

The work done against friction is:

$$Work = F \times s$$
$$Work = F \times 24$$

Substituting the expression for F found above gives:

Work done against friction $\quad = (0.25 \times 5\,\text{g}\cos 40) \times 24$
$$= 225.2$$

The work done against friction is 225 J.

(b) Work done against gravity = weight × vertical distance moved

$$= 5\,\text{g} \times 24 \sin 40$$
$$= 5 \times 9.8 \times 24 \sin 40$$
$$= 755.9$$

The work done against gravity is 756 J.

Forces at an angle to the direction of motion

Consider a particle P resting on a horizontal surface. If a force of magnitude F inclined at an angle θ to the horizontal causes the particle P to move along the surface while remaining in contact with the surface, you can resolve the force into its horizontal and vertical components.

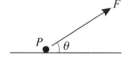

The vertical component does no work because the particle is moving in the horizontal direction only. For the horizontal component using:

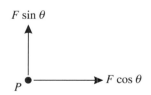

$$\text{Work done} = \text{force} \times \text{distance moved}$$

Gives: $\quad$ Work done $= F\cos\theta \times$ distance moved

■ **For a force at an angle to the direction of motion:**
Work done = component of force in direction of motion ×
distance moved in the same direction.

Example 19

A packing case is pulled across a smooth horizontal floor by a force
of magnitude 20 N inclined at 30° to the horizontal. Find the work
done by the force in moving the packing case a distance of 10 m.

Model the packing case as a particle moving horizontally.

Work done = horizontal component of force × distance moved

$$= 20 \cos 30 \times 10 = 173.2$$

The work done is 173 J.

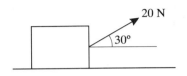

Exercise 5E

Whenever a numerical value of g is required take $g = 9.8 \, \mathrm{m \, s^{-2}}$.

1 Find the work done by a horizontal force of magnitude 0.7 N
   which pushes a particle a distance of 1.2 m across a horizontal
   surface.

2 A packing case is pushed 10 m across a horizontal floor by a
   constant horizontal force. If the work done by the force is
   55 J find the magnitude of the force.

3 A package of mass 20 kg is raised 3 m vertically. By modelling
   the package as a particle find the work done against gravity.

4 Find the work done against gravity when a particle of mass
   2 kg is raised a vertical distance of 8 m.

5 A packing case is pulled 12 m across a horizontal floor at a
   constant speed against resistances totalling 25 N. By
   modelling the packing case as a particle find the work done.

6 A cable is attached to a load of bricks of total mass 200 kg. If
   the work done in raising the bricks vertically at a constant
   speed from the ground to the top of a building is 23.5 kJ find
   the height of the building.

7 A packing case is pulled 20 m across a frozen pond by a rope
   inclined at 35° to the horizontal. The tension in the rope is
   25 N. By modelling the packing case as a particle and the
   surface of the pond as a smooth horizontal surface, calculate
   the work done.

8 A particle of mass 5 kg is pulled across a rough horizontal surface. The coefficient of friction between the body and the surface is 0.4. Given that the particle moves with speed 7 m s⁻¹, find the work done against friction in 2 seconds.

9 A body of mass 5 kg is pulled a distance of 12 m across a rough horizontal surface. The body moves with constant speed and the work done against friction is 147 J. Find the coefficient of friction between the body and the surface.

10 A climber of mass 70 kg climbs a vertical cliff of height 35 m. By modelling the climber as a particle find the work he does against gravity.

11 A body of mass 2 kg is pulled at a constant speed a distance of 20 m up a smooth plane inclined at 30° to the horizontal. Find the work done against gravity.

12 A boy of mass 30 kg slides 3 m down a slide inclined at 40° to the horizontal. By modelling the boy as a particle find the work done by gravity.

13 A particle of mass 6 kg is pulled at a constant speed a distance of 15 m up a rough surface inclined at 30° to the horizontal. The coefficient of friction between the particle and the surface is 0.2. Assuming the particle moves up a line of greatest slope find the work done against friction and the work done against gravity.

14 A rough surface is inclined at $\arctan \frac{3}{4}$ to the horizontal. A particle of mass 60 kg is pulled at a constant speed a distance of 40 m up a line of greatest slope of the surface. The coefficient of friction between the particle and the surface is $\frac{1}{4}$. Find (a) the magnitude of the frictional force acting on the particle (b) the work done against friction (c) the work done against gravity.

15 A rough surface is inclined at $\arcsin \frac{3}{5}$ to the horizontal. A particle of mass 5 kg is pulled at a constant speed a distance of 15 m up the surface by a force acting along a line of greatest slope. The coefficient of friction between the particle and the surface is $\frac{1}{3}$. Find the total work done, assuming the only resistances to motion are due to gravity and friction.

16 A rough surface is inclined at arctan $\frac{5}{12}$ to the horizontal. A particle of mass 10 kg is pulled at a constant speed a distance of 5 m up the surface by a force acting along a line of greatest slope. The only resistances to motion are those due to gravity and friction. Given that the work done by the force is 200 J find the coefficient of friction between the particle and the surface.

Energy

The energy a particle possesses is a measure of its capability to do work. Energy exists in various forms but in this book, only kinetic energy and potential energy will be considered.

Kinetic energy

The kinetic energy of a particle is the energy it possesses by virtue of its motion.

Consider a particle of mass m on a smooth horizontal surface. A constant horizontal force F applied to the particle moves it through a distance s and increases its speed from u to v.

$$\text{Work done on the particle} = F \times s$$

The equation of motion states that $F = mass \times acceleration$.

Using the constant acceleration formula $v^2 = u^2 + 2as$ gives:

$$a = \frac{v^2 - u^2}{2s}$$

Substituting a into the equation of motion gives:

$$F = \frac{m(v^2 - u^2)}{2s}$$

So:
$$Fs = \tfrac{1}{2}mv^2 - \tfrac{1}{2}mu^2$$

When a particle is in motion, the expression $\frac{1}{2}mass \times (velocity)^2$ is called the **kinetic energy** (or K.E.) of the particle.

■
$$\textbf{K.E.} = \tfrac{1}{2}mv^2$$

Since work done $= \tfrac{1}{2}mv^2 - \tfrac{1}{2}mu^2$:

$$\text{work done by a force} = \text{K.E. at end} - \text{K.E. at start}$$

or:
$$\text{work done} = \text{increase in K.E.}$$

If more than one force acts on the particle, for example a particle being pulled along a rough surface, then the force used to calculate the work done is the resultant force in the direction of the motion.

Because work done = increase in K.E., work and kinetic energy are measured in the same units.

For a mass in kilograms and a velocity in metres per second the kinetic energy is measured in joules.

Example 20
A particle of mass 0.5 kg is moving with a speed of $6\,\mathrm{m\,s^{-1}}$. Find its kinetic energy.

You know that:
$$\mathrm{K.E.} = \frac{1}{2}mv^2$$

Substituting known quantities gives:
$$\mathrm{K.E.} = \frac{1}{2} \times 0.5 \times 6^2$$
$$= 9$$

The kinetic energy of the particle is 9 J.

Example 21
A particle of mass 2 kg is being pulled across a smooth horizontal surface by a horizontal force. The force does 24 J of work in increasing the particle's velocity from $5\,\mathrm{m\,s^{-1}}$ to $v\,\mathrm{m\,s^{-1}}$. Find the value of v.

$$\text{Work done} = \tfrac{1}{2}mv^2 - \tfrac{1}{2}mu^2$$

Initial K.E. of particle is:

So:
$$\tfrac{1}{2}mu^2 = \tfrac{1}{2} \times 2 \times 5^2$$
$$24 = \tfrac{1}{2}mv^2 - \tfrac{1}{2} \times 2 \times 5^2$$
$$\tfrac{1}{2}mv^2 = 24 + 25$$
$$\tfrac{1}{2} \times 2v^2 = 49$$
$$v^2 = 49$$
$$v = 7$$

The final speed of the particle is $7\,\mathrm{m\,s^{-1}}$.

Example 22
A car of mass 1200 kg starts from rest at a set of traffic lights. After travelling 300 m its speed is $20\,\mathrm{m\,s^{-1}}$. Given that the car is subject to a constant resistance of 400 N find the driving force, assumed constant.

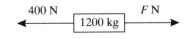

Increase in K.E. $= \frac{1}{2}mv^2 - \frac{1}{2}mu^2$

$$= \frac{1}{2} \times 1200 \times 20^2 - 0$$

Resultant force in direction of motion $= F - 400$.

Because: Work done $=$ increase in K.E.

and: Work done $=$ force $\times$ distance moved

Then: $(F - 400) \times 300 = \frac{1}{2} \times 1200 \times 20^2$

$$F - 400 = \frac{\frac{1}{2} \times 1200 \times 20^2}{300}$$

$$F = 800 + 400 = 1200$$

The driving force is 1200 N.

Potential energy

The **potential energy** of a particle is the energy it possesses by virtue of its position. As shown earlier on page 145, when a particle of mass m kg is *raised* through a vertical distance h metres the work done against gravity is mgh joules. This work is equal to the *increase* in potential energy (or P.E.) of the particle. If the particle is *lowered* through a vertical distance h' then its potential energy is *decreased* by mgh' joules. For each situation choose a zero level for P.E. and indicate this on the diagram. Changes in potential energy can then be calculated relative to that level.

∎   **P.E.** $= mgh$

Example 23

A parcel of mass 5 kg is raised vertically through a distance of 2 m. Find the increase in potential energy.

Model the parcel as a particle moving vertically upwards and take the original level of the parcel as the zero level for P.E.

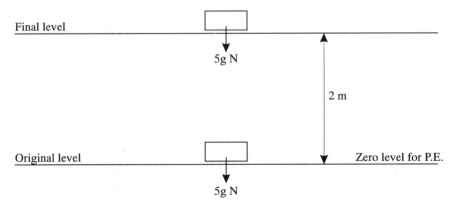

Initial P.E. $= 0$

Final P.E. $= mgh = 5 \times 9.8 \times 2 = 98$

The increase in P.E. is 98 J.

Example 24

A child of mass 25 kg slides 4 m down a playground slide inclined at an angle of arcsin $\frac{3}{5}$ to the horizontal. Model the child as a particle and the slide as an inclined plane and hence calculate the potential energy lost by the child.

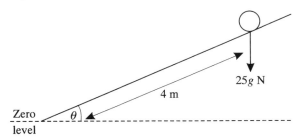

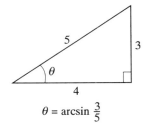

$\theta = \arcsin \frac{3}{5}$

Take the bottom of the slide to be the zero level for P.E.

Vertical distance moved by the child $= 4 \sin \theta = 4 \times \frac{3}{5}$

Initial P.E. $= mgh = 25 \times 9.8 \times 4 \times \frac{3}{5} = 588$

Final P.E. $= 0$

The potential energy lost by the child is 588 J.

Exercise 5F

1 Find the kinetic energy of:
   (a) a particle of mass 0.2 kg moving at $10\,\text{m s}^{-1}$
   (b) a particle of mass 2 kg moving at $2\,\text{m s}^{-1}$
   (c) a package of mass 10 kg moving at $5\,\text{m s}^{-1}$
   (d) a bullet of mass 15 g moving at $500\,\text{m s}^{-1}$
   (e) a car of mass 750 kg moving at $17\,\text{m s}^{-1}$
   (f) a man of mass 65 kg running at $5\,\text{m s}^{-1}$.

2 Find the change in potential energy of each of the following, stating in each case whether it is a loss or a gain:
   (a) a particle of mass 0.5 kg raised through a vertical distance of 7 m
   (b) a man of mass 65 kg descending a vertical distance of 20 m

- •(c) a lift of mass 600 kg ascending a vertical distance of 40 m
  - (d) a man of mass 70 kg ascending a vertical distance of 100 m
  - (e) a bucket of water of mass 10 kg raised a vertical distance of 22 m
  - (f) a lift of mass 900 kg descending a vertical distance of 30 m.

3 A particle of mass 0.5 kg increases its speed from 5 m s^{-1} to 7 m s^{-1}. Find the increase in the kinetic energy of the particle.

4 A car of mass 750 kg decreases speed from 30 m s^{-1} to 5 m s^{-1}. Find the decrease in the kinetic energy of the car.

5 A particle of mass 0.5 kg increases its kinetic energy by 15 J. Its initial speed was 2 m s^{-1}. Find its final speed.

6 A boy of mass 40 kg initially running at 4 m s^{-1} decreases his kinetic energy by 140 J. Find his final speed.

7 A child of mass 30 kg slides 5 m down a playground slide inclined at 40° to the horizontal. By modelling the child as a particle calculate the potential energy lost by the child.

8 A particle of mass 5 kg is moving at 2 m s^{-1} when it is acted on by a force of magnitude 10 N for 2 seconds, causing it to accelerate. Find the increase in kinetic energy.

9 A car of mass 750 kg initially travelling at 40 m s^{-1} retards at 2 m s^{-2} for 4 s. By modelling the car as a particle calculate the kinetic energy lost.

10 A stone of mass 0.8 kg is dropped from a height of 1.5 m into a pond. Find the kinetic energy of the stone as it hits the surface of the water. The stone begins to sink in the water with a speed of 2.1 m s^{-1}. Determine the kinetic energy lost when the stone strikes the water.

Conservation of energy

The only force acting on a particle of mass m falling freely under gravity is its weight. If the particle increases its speed from u m s^{-1} to v m s^{-1} while descending a distance h metres then:

$$\text{Work done by gravity} = mgh$$

And: Increase in K.E. of particle = final K.E. − initial K.E.

$$= \tfrac{1}{2}mv^2 - \tfrac{1}{2}mu^2$$

Since: Work done = increase in K.E.

It follows that: $mgh = \tfrac{1}{2}mv^2 - \tfrac{1}{2}mu^2$

But as seen earlier the work done by gravity, mgh, is also equal to the decrease in potential energy.

        mgh = decrease in P.E. = initial P.E. − final P.E.

So: decrease in P.E. = increase in K.E.

or: initial P.E. − final P.E. = final K.E. − initial K.E.

And so: initial (P.E. + K.E.) = final (K.E. + P.E.)

■ **A particle's total energy is constant, if it is subject only to gravity.**

This is true whether the particle moves vertically or along a path inclined to the vertical and is an application of the **Principle of conservation of energy**.

Example 25

A particle of mass 2 kg is released from rest and slides down a smooth plane inclined at $\arcsin \tfrac{3}{5}$ to the horizontal. Find the distance travelled while the particle increases its velocity to $5\,\mathrm{m\,s}^{-1}$.
Let the distance travelled be x m.

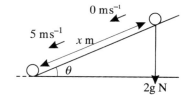

Then: Decrease in height $= x \sin \theta$

$$= \tfrac{3}{5}x$$

        Decrease in P.E. $= mgh$

$$= 2 \times 9.8 \times \tfrac{3}{5}x$$

      Increase in K.E. $= \tfrac{1}{2}mv^2 - \tfrac{1}{2}mu^2$

$$= \tfrac{1}{2} \times 2 \times 5^2 - 0$$

Since decrease in P.E. = increase in K.E:

$$2 \times 9.8 \times \tfrac{3}{5}x = \tfrac{1}{2} \times 2 \times 5^2$$

$$x = \frac{5 \times 5^2}{2 \times 9.8 \times 3} = 2.125$$

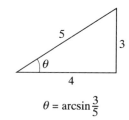

$\theta = \arcsin \tfrac{3}{5}$

The distance travelled is 2.13 m.

Change of energy of a particle

If the energy possessed by a particle changes during the motion being considered it must be as a result of some force doing work on that particle.

If a resistance is acting on the particle, the resistance is working against the motion and will cause a loss of energy.

Example 26

A particle of mass 5 kg is projected up a rough plane inclined at arcsin $\frac{4}{5}$ to the horizontal with a speed of 6 m s^{-1}. Given that the coefficient of friction between the particle and the plane is $\frac{1}{3}$ find the distance the particle moves up the plane before coming to rest.

Let the distance travelled up the plane be x metres.

Let FN be the friction force and RN the normal reaction.

The particle comes to rest because of the work done by friction.

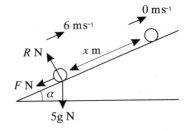

Initial K.E. $= \frac{1}{2}mv^2 = \frac{1}{2} \times 5 \times 6^2 = 90$

Final K.E. $= 0$

P.E. gained
$$\begin{aligned} &= mgh \\ &= 5 \times 9.8 \times x \sin \alpha \\ &= 5 \times 9.8 \times \tfrac{4}{5}x \\ &= 39.2x \end{aligned}$$

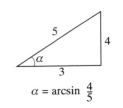

$\alpha = \arcsin \frac{4}{5}$

Total loss of energy $=$ K.E. lost $-$ P.E. gained

$$= (90 - 39.2x)$$

So total loss of energy is $(90 - 39.2x)\,$J

Work done by friction $= F \times x$

Since there is no motion perpendicular to the plane:

$$R - 5g\cos \alpha = 0$$

Therefore:
$$R = 5g\cos \alpha = 5g \times \tfrac{3}{5} = 3g$$

Because the particle is moving up the plane $F = \mu R$.

Therefore:
$$F = \tfrac{1}{3} \times 3g = g = 9.8$$

So work done by friction is $(9.8x)\,$J.

Since: Work done by friction = loss of energy

$$9.8x = 90 - 39.2x$$

$$x = \frac{90}{49} = 1.84$$

The distance moved is 1.84 m.

The above example can also be solved by using the equation of motion and the uniform acceleration equations. However, an examination question may specify 'by considering the energy of the particle' or similar wording in which case the above method must be used.

Example 27

A cyclist reaches the top of a hill with a speed of $4\,\mathrm{m\,s^{-1}}$. He descends 40 m and then ascends 35 m to the top of the next incline. His speed is now $3\,\mathrm{m\,s^{-1}}$. The cyclist and his bicycle have a combined mass of 90 kg. The total distance he cycles from the top of the first hill to the top of the next incline is 750 m and there is a constant resistance to motion of 15 N. Find the work done by the cyclist.

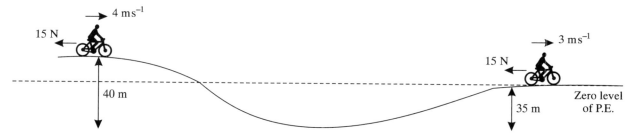

Take the cyclist's final level as the zero level for P.E.

Loss of K.E.
$$= \tfrac{1}{2}mu^2 - \tfrac{1}{2}mv^2$$
$$= \left(\tfrac{1}{2} \times 90 \times 4^2 - \tfrac{1}{2} \times 90 \times 3^2\right)$$
$$= 315$$

Loss of P.E.
$$= mgh$$
$$= 90 \times 9.8 \times (40 - 35)$$
$$= 4410$$

Total loss of energy is 4725 J.

Work done against resistances $F \times s = (15 \times 750) = 11250\,\mathrm{J}$.

The cyclist provides the work done against the resistances but he is 'helped' by his own loss of energy.

Work done by the cyclist = work done against the resistances −
energy lost

$$= (11250 - 4725) = 6525$$

Thus work done by the cyclist is 6525 J.

Exercise 5G

1 A particle of mass 0.6 kg falls a vertical distance of 5 m from rest. Neglecting resistances, find the potential energy lost and hence the final speed of the particle.

2 A stone of mass 1 kg is dropped from the top of a vertical cliff. If it hits the ground with a speed of $20 \, \text{m s}^{-1}$ find (a) the kinetic energy gained (b) the potential energy lost (c) the height of the cliff.

3 A package of mass 8 kg is pushed in a straight line across a smooth horizontal floor by a constant horizontal force of magnitude 16 N. The package has speed $3 \, \text{m s}^{-1}$ when it passes through point A and speed $5 \, \text{m s}^{-1}$ when it reaches point B. Find (a) the increase in kinetic energy of the package (b) the work done by the force (c) the distance AB.

4 A package of mass 6 kg is moving in a straight line across a rough horizontal surface. The coefficient of friction between the package and the surface is $\frac{1}{3}$. The package passes through point A with a speed of $5 \, \text{m s}^{-1}$ and comes to rest at B. Find (a) the kinetic energy lost by the package (b) the work done against friction (c) the distance AB.

5 A particle of mass 0.2 kg slides in a straight line across a rough horizontal surface. The speed of the particle decreases from $10 \, \text{m s}^{-1}$ to $5 \, \text{m s}^{-1}$ while travelling 9 m. Find (a) the kinetic energy lost (b) the work done against friction (c) the coefficient of friction between the particle and the surface.

6 A ball of mass 0.5 kg falls a vertical distance of 7 m from rest. By considering energy, find the final speed of the ball.

7 A stone of mass 0.2 kg is dropped by a bird which is flying horizontally. If the stone hits the horizontal ground with a

speed of $40 \, \text{m s}^{-1}$ by considering energy, find how high above the ground the bird is flying.

8 A bullet of mass $10 \, \text{g}$ travelling at $400 \, \text{m s}^{-1}$ horizontally hits a vertical wall and penetrates the wall to a depth of $40 \, \text{mm}$. Assuming the resistive force exerted by the wall to be constant find the magnitude of the force.

9 A bullet of mass $20 \, \text{g}$ travelling at $500 \, \text{m s}^{-1}$ horizontally hits a vertical wall. The wall exerts a constant resistance of $36\,000 \, \text{N}$ on the bullet. Calculate the distance the bullet will penetrate the wall.

10 A body of mass $3 \, \text{kg}$ is released from rest and slides $2 \, \text{m}$ down a smooth plane inclined at $40°$ to the horizontal. Find (a) the potential energy lost (b) the kinetic energy gained (c) the final speed of the body.

11 A body of mass $5 \, \text{kg}$ is released from rest and slides down a smooth plane inclined at $30°$ to the horizontal. Find the distance travelled while increasing its speed to $4 \, \text{m s}^{-1}$.

12 A particle of mass $0.4 \, \text{kg}$ is projected up a line of greatest slope of a smooth plane inclined at $35°$ to the horizontal with a velocity of $12 \, \text{m s}^{-1}$. Use energy considerations to find the distance the particle travels before first coming to rest.

13 A particle of mass $0.5 \, \text{kg}$ is projected up a line of greatest slope of a smooth plane inclined at $\arcsin \frac{3}{5}$ to the horizontal. Given that the particle travels a distance of $10 \, \text{m}$ before first coming to rest find the speed of projection.

14 A particle of mass $4 \, \text{kg}$ is projected up a rough plane inclined at $\arcsin \frac{5}{13}$ to the horizontal with a velocity of $8 \, \text{m s}^{-1}$. Given that the coefficient of friction between the single particle and the plane is $\frac{1}{4}$ find, by energy considerations, the distance the particle moves up the plane before coming to rest.

15 A cyclist and his machine have a combined mass of $90 \, \text{kg}$. The cyclist freewheels down a hill inclined at $30°$ to the horizontal. He increases his speed from $5 \, \text{m s}^{-1}$ to $25 \, \text{m s}^{-1}$. By modelling the cyclist and his bicycle as a single particle and assuming resistances can be neglected, calculate the potential energy lost by the cyclist and the distance he travels.

16 A cyclist starts from rest and freewheels down a hill inclined at arcsin $\frac{1}{20}$ to the horizontal. After travelling 80 m the road becomes horizontal and the cyclist travels another 80 m before coming to rest without using his brakes. Given that the combined mass of the cyclist and his machine is 80 kg, model the cyclist and his machine as a single particle and hence find the resistive force, assumed constant throughout the motion.

17 A boy and his skateboard have a mass of 50 kg. He descends a slope inclined at 12° to the horizontal starting from rest. At the bottom, the ground becomes horizontal for 10 m before rising at 8° to the horizontal. The boy travels 30 m up the incline before coming to rest again. He is subject to a constant resistance of 20 N throughout the motion. By modelling the boy and his skateboard as a single particle find the distance the boy travelled down the slope.

18 A tile slides down a smooth roof inclined at 45° to the horizontal and off the edge which is 10 m above the ground. Given that the tile started from rest 5 m from the edge of the roof find the velocity with which it hits the ground.

19 A book slides down a rough desk lid which is inclined at 30° to the horizontal and falls to the ground which is 0.8 m below the edge of the lid. The coefficient of friction between the book and the lid is $\frac{1}{4}$. Given that the book starts from rest 0.4 m from the bottom of the lid which is hinged at the edge of the desk, find the velocity with which the book hits the floor.

20 A rough plane is inclined at arcsin $\frac{3}{5}$ to the horizontal. A package of mass 40 kg is released from rest and travels 15 m while increasing speed to 12 m s^{-1}. Find, using energy considerations, the coefficient of friction between the package and the plane.

Power

Power is defined as the rate of doing work. It is measured in watts (W) where **1 watt is 1 joule per second**. The power of an engine is frequently given in kilowatts (kW). **1 kilowatt is 1000 watts**. An engine which develops a power of 1 kilowatt is doing 1000 joules of work every second.

Example 28

A force of magnitude 1500 N pulls a truck up a slope at a constant speed of $6 \, \text{m s}^{-1}$. Given that the force acts parallel to the direction of motion find in kW the power developed.

Work done per second = Force × distance moved in 1 second

$$= (1500 \times 6) = 9000 \, \text{J}$$

Power = rate of doing work = $9000 \, \text{W}$

The power developed is 9 kW.

Moving vehicles

If the engine of a vehicle is producing a driving force of F newtons when the vehicle has a speed of $v \, \text{m s}^{-1}$ then the work done per second is $F \times$ distance moved in one second $= F \times v$.

Because work done per second is the power developed by the engine:

■   **Power** $= F \times v$

Example 29

A car of mass 1000 kg is travelling along a level road against a constant resistance of magnitude 475 N. The engine of the car is working at 4 kW. Calculate:
(a) the acceleration when the car is travelling at $5 \, \text{m s}^{-1}$
(b) the maximum speed of the car.

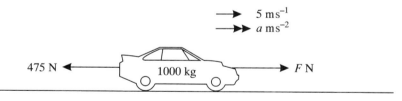

(a) Power = 4 kW = 4000 W

Since: Power $= F \times v$

So: $4000 = F \times 5$

$$F = 800$$

Using $F = ma$ where F is the resultant force gives:

$$800 - 475 = 1000a$$

$$a = \frac{800 - 475}{1000} = 0.325$$

The acceleration is $0.325\,\text{m s}^{-2}$.

(b) When the car is travelling at its maximum speed there will be no acceleration so the resultant force in the direction of motion will be zero.

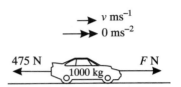

Hence: $\qquad\qquad\qquad F = 475N$

Using: $\qquad\qquad\qquad \text{Power} = F \times v$

Gives: $\qquad\qquad\qquad 4000 = 475v$

$$v = \frac{4000}{475} = 8.421$$

The maximum velocity of the car is $8.42\,\text{m s}^{-1}$.

Example 30

A car of mass 1200 kg is moving up a hill of slope arc sin $\frac{1}{15}$ at a constant speed of $20\,\text{m s}^{-1}$. If the power developed by the engine is 25 kW find the resistance to motion.

At the top of the hill the road becomes horizontal. Find the initial acceleration, assuming the resistance to be unchanged.

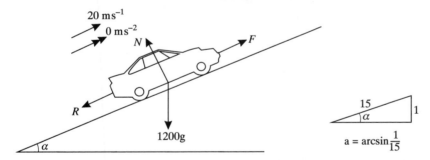

The driving force produced by an engine is often called the **tractive force.**

Let the tractive force be F.

Because: $\qquad\qquad\qquad \text{Power} = F \times v$

$$25 \times 10^3 = F \times 20$$

$$F = \frac{25 \times 10^3}{20}$$

$$= 1250$$

Since the speed is constant, there is no acceleration, and so there is no resultant force in the direction of motion. Hence:

$$R + 1200\,g \sin \theta = F$$
$$= 1250$$

So:
$$R = 1250 - 1200 \times 9.8 \times \frac{1}{15}$$
$$= 466$$

The resistance to motion is 466 N.

At the instant the road becomes horizontal, the tractive force is 1250 N.

Using:
$$F = ma$$

Gives:
$$1250 - 466 = 1200a$$
$$a = \frac{1250 - 466}{1200} = 0.6533$$

The initial acceleration when the car travels along the horizontal part of the road is $0.653\,\mathrm{m\,s^{-2}}$.

In the above example, as the car's speed increases either the power which equals force × velocity, must increase or the driving force must decrease and there will be a corresponding decrease in acceleration.

Exercise 5H

Whenever a numerical value of g is required take $g = 9.8\,\mathrm{m\,s^{-2}}$.

1 A force of 1200 N pulls a truck up a slope at a constant speed of $10\,\mathrm{m\,s^{-1}}$. Find in kW the power developed.

2 The engine of a car is producing a driving force of 800 N. Given that the car is travelling at $12\,\mathrm{m\,s^{-1}}$ find the power developed.

3 The engine of a car is working at 4 kW. Given that the car is travelling at $15\,\mathrm{m\,s^{-1}}$ find the driving force produced by the engine.

4 A car is travelling along a level road against a constant resistance of 500 N. The engine of the car is working at 4.5 kW. Calculate the maximum speed of the car.

5 A car is travelling along a level road against a constant resistance of 475 N. The engine of the car is working at 6 kW. Calculate the maximum speed of the car.

6 A car is travelling along a level road at a constant speed of $15\,\text{m}\,\text{s}^{-1}$. The engine of the car is working at 7.5 kW. Calculate the resistance to motion.

7 A car of mass 900 kg is travelling along a level road against a constant resistance of 400 N. The engine of the car is working at 8 kW. Calculate (a) the acceleration when the car is travelling at $5\,\text{m}\,\text{s}^{-1}$ (b) the acceleration when the car is travelling at $12\,\text{m}\,\text{s}^{-1}$ (c) the maximum speed of the car.

8 A car of mass 1000 kg is travelling along a level road against a constant resistance. When the engine is working at 12 kW and the speed of the car is $15\,\text{m}\,\text{s}^{-1}$, the acceleration is $0.4\,\text{m}\,\text{s}^{-2}$. Find the resistive force.

9 A car of mass 800 kg is travelling along a level road at a speed of $20\,\text{m}\,\text{s}^{-1}$. The car is subject to a constant resistance of 250 N and its acceleration is $0.25\,\text{m}\,\text{s}^{-2}$. Find the power developed by the engine.

10 A cyclist is travelling along a level road against a constant resistance of 30 N. The maximum rate at which the cyclist can work is 300 W. Find her maximum speed.

11 A car of mass 1000 kg is moving up a hill inclined at arcsin $\frac{1}{20}$ at a steady speed of $25\,\text{m}\,\text{s}^{-1}$. Given that the power developed by the engine is 20 kW find the resistance to motion. The road now becomes horizontal. Find the initial acceleration, assuming the resistance to be unchanged.

12 A car of mass 1000 kg is travelling along a level road against a constant resistance of 400 N. Given that the maximum speed of the car is $35\,\text{m}\,\text{s}^{-1}$ find the power developed by the engine. With the engine of the car working at the same rate and the resistance unchanged the car ascends a slope inclined at 10° to the horizontal. Find its maximum velocity up the hill.

13 The mass of a car is 750 kg. Its engine is working at a constant rate of 15 kW against a constant resistance of magnitude 600 N. Given that the car is travelling at $20\,\text{m}\,\text{s}^{-1}$

along a horizontal road find the acceleration of the car. The car now ascends a straight road, inclined at 6° to the horizontal. Given that the car's engine works at the same rate and the resistance to motion is unchanged find the maximum speed of the car up the slope.

14 A train of mass 200 tonnes is ascending a hill inclined at 0.5° to the horizontal. The engine is working at a constant rate of 300 kW and the resistances to motion amount to 8 kW. Find the maximum speed of the train. The track now becomes horizontal. Given that the engine works at the same rate and the resistances to motion are unchanged find the initial acceleration of the train.

15 The resistance to motion of a car moving with speed $v\,\mathrm{m\,s^{-1}}$ is given by the formula $(200 + 2v)\,\mathrm{N}$. Given that the engine of the car is working at 8 kW find the maximum speed of the car as it travels along a horizontal road.

5.4 Momentum and impulse

Suppose a particle of mass m moving in a straight line with constant acceleration increases its speed from u to v in time t. Then its acceleration can be found by using:

$$v = u + at$$

Or:

$$a = \frac{v - u}{t}$$

Applying the equation of motion to find the force required to produce this acceleration gives:

$$F = ma = m\frac{(v - u)}{t}$$

Or:

$$Ft = mv - mu \tag{1}$$

Momentum

For any moving particle, the quantity mass × velocity is called the **momentum** of the particle. Momentum is a vector quantity because it depends on the particle's velocity.

Impulse

If a constant force **F** is applied to a particle for a period of time t, then the quantity force × time ($F \times t$) is called the **impulse**, **I**, of the force on the particle. Equation (1) can now be re-written as:

$$\mathbf{I} = m\mathbf{v} - m\mathbf{u}$$

where **u** is the initial velocity and **v** is the final velocity.

Or: Impulse = change in momentum.

For a mass in kg and a velocity in $\mathrm{m\,s^{-1}}$ force would be in newtons and hence impulse and momentum in newton-seconds (Ns).

Example 31

A particle of mass 5 kg is at rest on a smooth horizontal surface. A horizontal force of magnitude 4 N acts on the particle for 6 seconds. Find the final speed of the particle.

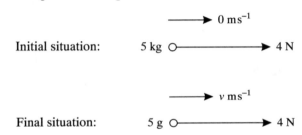

Let the final speed be $v\,\mathrm{m\,s^{-1}}$. This will be in the same direction as the applied force.

Using: $Ft = mv - mu$

$$4 \times 6 = 5v - 0$$
$$v = 4.8$$

The final speed of the particle is $4.8\,\mathrm{m\,s^{-1}}$.

In some situations, for example when a cricketer hits the ball, it is not possible to measure the length of time for which the bodies are in contact. Only the impulse of the force between the two bodies can be calculated.

Example 32

A ball of mass $\frac{1}{4}$ kg hits a vertical wall with a horizontal speed of $30\,\mathrm{m\,s^{-1}}$. It rebounds with a speed of $20\,\mathrm{m\,s^{-1}}$. Find the impulse exerted by the wall on the ball.

In problems of this type it is helpful to draw diagrams to show the situation before and after the impact.

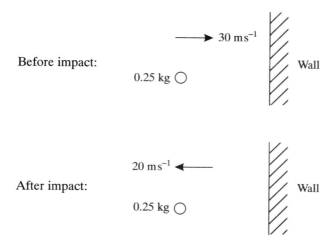

Before impact:

30 ms^{-1}

0.25 kg ◯

Wall

After impact:

20 ms^{-1}

0.25 kg ◯

Wall

Because the velocity arrows in the two diagrams are in opposite directions, different signs must be used when substituting these velocities in the impulse-momentum equation.

Using: $$I = mv - mu$$

Taking the direction away from the wall to be positive gives:

$$I = 0.25 \times 20 - 0.25 \times (-30)$$
$$I = 12.5$$

The impulse exerted on the ball is 12.5 Ns.

By Newton's third law the force, and hence the impulse, on the ball is equal in magnitude to the force, or impulse, exerted on the wall. So the ball exerts an impulse of magnitude 12.5 Ns on the wall.

As momentum is a vector quantity, questions involving momentum may be expressed in vector form.

Example 33

At time $t = 0$ a particle of mass 8 kg is moving with a velocity of $(2\mathbf{i} + 2\mathbf{j})\,\mathrm{m\,s}^{-1}$. Four seconds later its velocity is $(6\mathbf{i} - 4\mathbf{j})\,\mathrm{m\,s}^{-1}$. Find the constant force acting on the particle.

Let the force be $(x\mathbf{i} + y\mathbf{j})\,\mathrm{N}$.

Using: $$\mathbf{F}t = m\mathbf{v} - m\mathbf{u}$$

Gives:
$$(x\mathbf{i} + y\mathbf{j}) \times 4 = 8(6\mathbf{i} - 4\mathbf{j}) - 8(2\mathbf{i} + 2\mathbf{j})$$
$$(x\mathbf{i} + y\mathbf{j}) = 2(6\mathbf{i} - 4\mathbf{j}) - 2(2\mathbf{i} + 2\mathbf{j})$$
$$x\mathbf{i} + y\mathbf{j} = 8\mathbf{i} - 12\mathbf{j}$$

The force is $(8\mathbf{i} - 12\mathbf{j})\,\mathrm{N}$.

Exercise 5I

1 Calculate the magnitude of the momentum of the following:
   (a) a particle of mass $0.02\,\text{kg}$ moving at $10\,\text{m}\,\text{s}^{-1}$
   (b) a particle of mass $200\,\text{g}$ moving at $100\,\text{m}\,\text{s}^{-1}$
   (c) a car of mass $900\,\text{kg}$ moving at $15\,\text{m}\,\text{s}^{-1}$
   (d) a man of mass $80\,\text{kg}$ running at $6\,\text{m}\,\text{s}^{-1}$
   (e) a train of mass 150 tonnes moving at $20\,\text{m}\,\text{s}^{-1}$
   (f) a car of mass $750\,\text{kg}$ moving at $18\,\text{m}\,\text{s}^{-1}$.

2 A car of mass $800\,\text{kg}$ decelerates from $20\,\text{m}\,\text{s}^{-1}$ to $5\,\text{m}\,\text{s}^{-1}$. Find the loss of momentum.

3 A car of mass $900\,\text{kg}$ accelerates from rest to $18\,\text{m}\,\text{s}^{-1}$. Find the gain in momentum.

4 A particle of mass $0.2\,\text{kg}$ is moving with speed $5\,\text{m}\,\text{s}^{-1}$. Its speed changes to $8\,\text{m}\,\text{s}^{-1}$ and its direction of travel is reversed. Find the change of momentum of the particle.

5 A toy car of mass $0.5\,\text{kg}$ is pushed across a smooth horizontal floor by a horizontal force of $3\,\text{N}$ for $4\,\text{s}$. Find the gain in momentum of the car. If the car starts from rest find its final speed.

6 A parcel of mass $10\,\text{kg}$ is at rest on a smooth horizontal surface. It is pulled by a horizontal force of $15\,\text{N}$ for 3 seconds. Model the car as a particle and hence find its final speed.

7 A particle of mass $6\,\text{kg}$ is at rest on a smooth horizontal surface. It is acted on by a constant horizontal force for $5\,\text{s}$ and then has a speed of $15\,\text{m}\,\text{s}^{-1}$. Find the magnitude of the force.

8 A particle of mass $10\,\text{kg}$ is moving with a velocity of $7\mathbf{i}\,\text{m}\,\text{s}^{-1}$ when it is acted on by a force of $3\mathbf{i}\,\text{N}$ for $2\,\text{s}$. Find (a) the impulse given to the particle by the force (b) the final velocity of the particle.

9 A particle of mass $4\,\text{kg}$ is moving with a velocity of $7\mathbf{j}\,\text{m}\,\text{s}^{-1}$; $3\,\text{s}$ later it has a velocity of $-5\mathbf{j}\,\text{m}\,\text{s}^{-1}$. Find the constant force acting on the particle.

10 A particle of mass 0.2 kg is moving with a velocity of $8\mathbf{i}\,\mathrm{m\,s^{-1}}$; t s later its velocity is $16\mathbf{i}\,\mathrm{m\,s^{-1}}$. If the constant force acting on the particle has magnitude 0.4 N find the value of t.

11 A particle of mass 6 kg is moving with a velocity of $(3\mathbf{i} - 2\mathbf{j})\,\mathrm{m\,s^{-1}}$; 3 s later it has a velocity of $(7\mathbf{i} + 3\mathbf{j})\,\mathrm{m\,s^{-1}}$. Find the constant force acting on the particle.

12 A ball of mass 0.2 kg hits a vertical wall with a horizontal speed of $15\,\mathrm{m\,s^{-1}}$. It rebounds with a speed of $12\,\mathrm{m\,s^{-1}}$. Find the impulse exerted on the ball.

13 A smooth sphere of mass 1.5 kg hits a vertical wall with a horizontal speed of $2\,\mathrm{m\,s^{-1}}$. It rebounds with a velocity of $1.5\,\mathrm{m\,s^{-1}}$. Find (a) the impulse exerted on the sphere (b) the impulse exerted on the wall.

14 A ball of mass 0.5 kg is lying on the ground. It receives an impulse of 15 Ns from a bat. Model the ball as a particle and hence find the velocity with which the ball moves.

15 A ball of mass 0.4 kg is dropped from a height of 2.5 m. After hitting the ground it rises to a height of 1.8 m. By modelling the ball as a particle find (a) the speed with which the ball hits the ground (b) the speed with which the ball rebounds from the ground (c) the impulse the ball receives from the ground.

The principle of conservation of momentum

When two particles collide, by Newton's third law they exert equal and opposite forces, and hence impulses, on each other.

Consider two particles of masses m_1 and m_2 moving in the same direction in a straight line on a smooth horizontal surface with speeds u_1 and u_2 respectively where $u_1 > u_2$. The particles collide. Let their speeds after impact be v_1 and v_2 respectively in the same direction as u_1 and u_2 and let the impulse created by the impact be I.

Before impact:

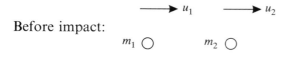

To show the impulse when the particles are in contact it is necessary to draw the particles a small distance apart.

During impact: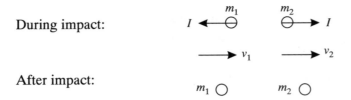

After impact:

Because: Impulse = change of momentum

for each particle in turn use:

$$I = mv - mu$$

Take left → right to be the positive direction:

for m_1: $-I = m_1 v_1 - m_1 u_1$ (1)

for m_2: $I = m_2 v_2 - m_2 u_2$ (2)

Adding equations (1) and (2) gives:

$$0 = m_1 v_1 - m_1 u_1 + m_2 v_2 - m_2 u_2$$

Rearranging gives:

$$m_1 u_1 + m_2 u_2 = m_1 v_1 + m_2 v_2$$

Or in words:

■ **Total momentum before impact = total momentum after impact**

This is known as the **principle of conservation of momentum**.

Example 34

Particle A of mass $2\,kg$ is moving with speed $3\,m\,s^{-1}$ on a smooth horizontal surface. Particle B is at rest on the surface. A collides with B. After impact A continues to move in the same direction but with a speed of $1\,m\,s^{-1}$. Find the speed of B.

Draw diagrams to show the situations before and after impact. Let the speed of B after impact be $v\,m\,s^{-1}$.

Before impact: $\longrightarrow 3\,ms^{-1}$ $\longrightarrow 0\,ms^{-1}$

A ◯ $2\,kg$ B ◯ $2\,kg$

After impact: $\longrightarrow 1\,ms^{-1}$ $\longrightarrow v\,ms^{-1}$

A ◯ $2\,kg$ B ◯ $3\,kg$

Take left → right to be the positive direction.

By the principle of conservation of momentum:

$$m_1u_1 + m_2u_2 = m_1v_1 + m_2v_2$$

So:
$$2 \times 3 + 3 \times 0 = 2 \times 1 + 3v$$
$$3v = 4$$
$$v = \tfrac{4}{3} = 1\tfrac{1}{3}$$

The speed of B after impact is $1\tfrac{1}{3}\,\mathrm{m\,s}^{-1}$.

Example 35

Two particles P and Q of masses $3\,\mathrm{kg}$ and $5\,\mathrm{kg}$ respectively are moving towards each other along the same straight line with speeds $4\,\mathrm{m\,s}^{-1}$ and $2\,\mathrm{m\,s}^{-1}$ respectively. After impact the direction of motion of P is reversed and its speed is $2\,\mathrm{m\,s}^{-1}$. Find the speed of Q.

Before impact:

$\longrightarrow 4\,\mathrm{ms}^{-1} \quad 2\,\mathrm{ms}^{-1}\longleftarrow$

$P \bigcirc 3\,\mathrm{kg} \qquad Q \bigcirc 5\,\mathrm{kg}$

After impact:

$2\,\mathrm{ms}^{-1}\longleftarrow \qquad \longrightarrow v\,\mathrm{ms}^{-1}$

$P \bigcirc 3\,\mathrm{kg} \qquad Q \bigcirc 5\,\mathrm{kg}$

Let the speed of Q after impact be $v\,\mathrm{m\,s}^{-1}$ as shown.

Take left $\rightarrow$ right to be the positive direction.

By the principle of conservation of momentum:

$$m_1u_1 + m_2u_2 = m_1v_1 + m_2v_2$$

So:
$$3 \times 4 - 5 \times 2 = -3 \times 2 + 5v$$
$$12 - 10 + 6 = 5v$$
$$v = \tfrac{8}{5} = 1\tfrac{3}{5}$$

The speed of Q after impact is $1\tfrac{3}{5}\,\mathrm{m\,s}^{-1}$.

Example 36

A snooker ball P moving with speed $4\,\mathrm{m\,s}^{-1}$ hits a stationary ball Q of equal mass. After the impact both balls move in the same direction along the same straight line, but the speed of Q is twice that of P. By modelling the balls as particles moving on a smooth horizontal surface find the speeds of the balls.

Let the balls have mass $m\,\mathrm{kg}$ and let the speed of P after impact be $v\,\mathrm{m\,s}^{-1}$. The speed of Q will be $2v\,\mathrm{m\,s}^{-1}$.

Before impact:

$\xrightarrow{\hspace{1cm}}$ 4 ms⁻¹ $\quad$ $\xrightarrow{\hspace{1cm}}$ 0 ms⁻¹

$P \bigcirc m$ kg $\qquad Q \bigcirc mg$

After impact:

$\xrightarrow{\hspace{1cm}}$ v ms⁻¹ $\quad$ $\xrightarrow{\hspace{1cm}}$ 2v ms⁻¹

$P \bigcirc m$ kg $\qquad Q \bigcirc mg$

Take left $\rightarrow$ right to be the positive direction.

By the principle of conservation of momentum:

$$m_1 u_1 + m_2 u_2 = m_1 v_1 + m_2 v_2$$

So:
$$4m + 0 = mv + 2mv$$
$$4 = 3v$$
$$v = \tfrac{4}{3} = 1\tfrac{1}{3}$$

After impact P has speed $1\tfrac{1}{3}$ m s⁻¹ and Q has speed $2\tfrac{2}{3}$ m s⁻¹.

Jerk in a string

The principle of conservation of momentum also applies to jerks in strings.

Consider two particles P and Q which are at rest on a smooth horizontal surface and are connected by a light inextensible string which is initially slack.

Suppose Q is given a velocity in the direction PQ. In time, the string will become taut. At the instant when the string becomes taut both particles will experience a jerk from the string.

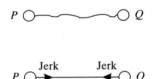

By Newton's third law, the jerk experienced by P will be equal in magnitude but opposite in direction to the jerk experienced by Q. Therefore momentum is conserved.

Example 37

Two particles P and Q of mass 3 kg and 6 kg respectively are connected by a light inextensible string. Initially they are at rest on a smooth table with the string slack. Q is projected directly away from P with a speed of 3 m s⁻¹. Find their common speed when the string becomes taut.

Let the common speed of P and Q after the jerk be $v\,\mathrm{m\,s}^{-1}$.

Before jerk: 3 kg $\bigcirc\!\!\overset{P}{\wedge\!\!\wedge\!\!\wedge}\overset{Q}{\bigcirc}$ 6 kg $\rightarrow 0\,\mathrm{ms}^{-1}$ $\rightarrow 3\,\mathrm{ms}^{-1}$

After jerk: $P\ \bigcirc\!\!\overset{3\,\mathrm{kg}}{\rule{2cm}{0.4pt}}\overset{Q}{\bigcirc}$ 6 kg $\rightarrow v\,\mathrm{ms}^{-1}$ $\rightarrow v\,\mathrm{ms}^{-1}$

Take left $\rightarrow$ right to be the positive direction.

By the principle of conservation of momentum:

$$m_1u_1 + m_2u_2 = m_1v_1 + m_2v_2$$

So:
$$3 \times 0 + 6 \times 3 = 3v + 6v$$
$$9v = 18$$
$$v = 2$$

The common speed of P and Q is $2\,\mathrm{m\,s}^{-1}$.

Energy changes resulting from impacts

When two particles collide there is no change in the momentum of the two particles but in most cases there is a change in their total energy.

For the particles in Example 35:

Initial K.E. $= \frac{1}{2}m_1u_1^2 + \frac{1}{2}m_2u_2^2$

$\qquad = (\frac{1}{2} \times 3 \times 16 + \frac{1}{2} \times 5 \times 4) = 34$

Final K.E. $= \frac{1}{2}m_1v_1^2 + \frac{1}{2}m_2v_2^2$

$\qquad = (\frac{1}{2} \times 3 \times 4 + \frac{1}{2} \times 5 \times \frac{64}{25}) = 12\frac{2}{5}$

Hence there is a loss of K.E. of $(34 - 12\frac{2}{5}) = 21\frac{3}{5}\,\mathrm{J}$.

In some situations, such as a gun firing a bullet, it is possible for the total K.E. to increase.

Example 38

A gun of mass 4 kg fires a bullet of mass 50 g. If the gun recoils with a speed of of $6\,\mathrm{m\,s}^{-1}$ find the speed, assumed horizontal, of the bullet and the gain in kinetic energy of the system.

Let the speed of the bullet be $v\,\mathrm{m\,s}^{-1}$.

Before firing the bullet:

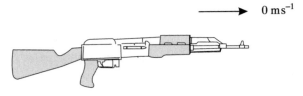

The gun and bullet are both at rest.

After firing bullet:

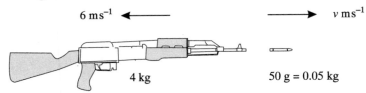

Take left $\rightarrow$ right to be the positive direction.

By the principle of conservation of momentum:

$$m_1u_1 + m_2u_2 = m_1v_1 + m_2v_2$$

So:
$$0 = 0.05v - 4 \times 6$$
$$v = \frac{4 \times 6}{0.05} = 480$$

The speed of the bullet is $480\,\mathrm{m\,s}^{-1}$.

Initial K.E. of system $= 0$ (as both gun and bullet are at rest)

Final K.E. of system $= \frac{1}{2}m_1v_1^2 + \frac{1}{2}m_2v_2^2$

$\qquad\qquad\qquad\quad = \frac{1}{2} \times 4 \times 6^2 + \frac{1}{2} \times 0.05 \times 480^2$

$\qquad\qquad\qquad\quad = 5832$

The K.E. gained is $5832\,\mathrm{J}$.

Exercise 5J

The diagrams for questions 1 to 10 show the collision between two particles A and B. Find the unknown mass m or velocity v in each case.

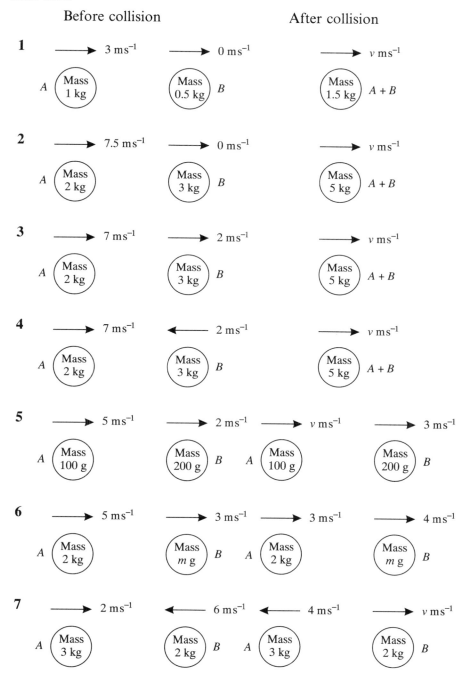

Before collision   After collision

1   $\longrightarrow$ 3 ms^{-1} $\longrightarrow$ 0 ms^{-1} $\longrightarrow$ v ms^{-1}

A (Mass 1 kg)   (Mass 0.5 kg) B   (Mass 1.5 kg) A + B

2   $\longrightarrow$ 7.5 ms^{-1} $\longrightarrow$ 0 ms^{-1} $\longrightarrow$ v ms^{-1}

A (Mass 2 kg)   (Mass 3 kg) B   (Mass 5 kg) A + B

3   $\longrightarrow$ 7 ms^{-1} $\longrightarrow$ 2 ms^{-1} $\longrightarrow$ v ms^{-1}

A (Mass 2 kg)   (Mass 3 kg) B   (Mass 5 kg) A + B

4   $\longrightarrow$ 7 ms^{-1} $\longleftarrow$ 2 ms^{-1} $\longrightarrow$ v ms^{-1}

A (Mass 2 kg)   (Mass 3 kg) B   (Mass 5 kg) A + B

5   $\longrightarrow$ 5 ms^{-1} $\longrightarrow$ 2 ms^{-1} $\longrightarrow$ v ms^{-1} $\longrightarrow$ 3 ms^{-1}

A (Mass 100 g)   (Mass 200 g) B   A (Mass 100 g)   (Mass 200 g) B

6   $\longrightarrow$ 5 ms^{-1} $\longrightarrow$ 3 ms^{-1} $\longrightarrow$ 3 ms^{-1} $\longrightarrow$ 4 ms^{-1}

A (Mass 2 kg)   (Mass m g) B   A (Mass 2 kg)   (Mass m g) B

7   $\longrightarrow$ 2 ms^{-1} $\longleftarrow$ 6 ms^{-1} $\longleftarrow$ 4 ms^{-1} $\longrightarrow$ v ms^{-1}

A (Mass 3 kg)   (Mass 2 kg) B   A (Mass 3 kg)   (Mass 2 kg) B

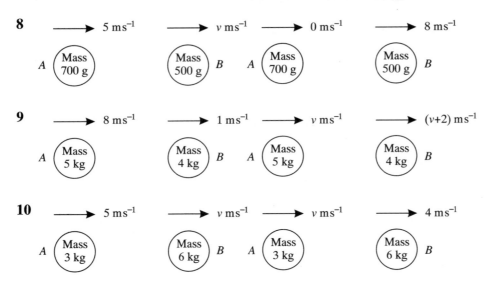

Before collision   After collision

11 A railway truck is travelling at $6\,\text{m s}^{-1}$ along a straight piece of track when it hits an identical truck which is stationary. The two trucks couple together automatically. By modelling the trucks as particles find their common speed after the collision.

12 A small smooth sphere A of mass $2.5\,\text{kg}$ lies at rest on a smooth horizontal table. A second small smooth sphere B of mass $1.5\,\text{kg}$ is moving with speed $4\,\text{m s}^{-1}$ and collides directly with A. The two spheres coalesce – in other words after impact they move as a single body. Find their speed after impact and the loss of kinetic energy of the system due to the impact.

13 Two particles P and Q of masses $3\,\text{kg}$ and $2\,\text{kg}$ respectively are at rest on a smooth horizontal surface. They are connected by a light inelastic string which is initially slack. P is projected away from Q with a speed of $10\,\text{m s}^{-1}$. Find the common speed of the particles after the string becomes taut and the impulse in the string when it jerks tight.

14 Particle A of mass $1\,\text{kg}$ is at rest $0.2\,\text{m}$ from the edge of a smooth horizontal $0.8\,\text{m}$ high table. It is connected by a light inextensible string of length $0.7\,\text{m}$ to a particle B of mass $0.5\,\text{kg}$. Particle B is initially at rest at the edge of the table closest to A but then falls off. Assuming B's initial horizontal velocity to be zero find the speed with which *A* begins to move.

15 A gun of mass 5 kg fires a bullet of mass 40 g. Given that the bullet leaves the gun with a speed of 500 m s^{-1} find the initial speed of recoil of the gun and the gain in kinetic energy of the system.

16 A gun of mass 5000 kg discharges a shot of mass 25 kg. The gun recoils against a constant force of 10 kN which brings it to rest in 1.25 s. Find the speed of the shot.

17 A shell is travelling horizontally at 400 m s^{-1} when it explodes into two pieces whose masses are in the ratio 2 : 3. The larger piece has a speed of 800 m s^{-1} in the original direction of motion. Given that the smaller piece also moves horizontally after the explosion find its velocity. Find also the increase in kinetic energy of the system as a multiple of the kinetic energy of the original shell.

18 A wooden peg of mass 0.4 kg is driven vertically into the ground by a mallet of mass 2 kg moving vertically downwards. The speed of the mallet just prior to impact is 8 m s^{-1}. After impact the mallet and peg remain in contact. By modelling the peg and mallet as particles find the speed with which the peg begins to enter the ground. If the ground offers a resistance to motion of 1 kN find how far the peg is driven into the ground due to the impact.

SUMMARY OF KEY POINTS

1 **Newton's Laws of Motion**
   1 A particle will only accelerate if it is acted on by a resultant force.
   2 The force F applied to a particle is proportional to the mass *m* of the particle and the acceleration produced.
$$F = ma$$
   3 The forces between two bodies in contact are equal in magnitude but opposite in direction.

2 **Work**

For a force acting in the direction of the motion:

Work done = Force × distance moved

For a force acting in a direction other than that of the motion:

Work done = component of force in the direction of motion × distance moved in the same direction

A force of 1 newton (N) does 1 joule (J) of work when moving a particle a distance of 1 metre.

3 Energy

The kinetic energy (K.E.) of a body of mass m moving with speed $v\,\mathrm{m}s^{-1}$ is given by:

$$\text{K.E.} = \tfrac{1}{2}mv^2$$

For a mass in kg and velocity in $\mathrm{m\,s^{-1}}$ the K.E. is measured in joules (J). K.E. is never negative.

The potential energy (P.E.) of a body of mass m at a height h above a chosen level is given by:

$$\text{P.E.} = mgh$$

Potential energy is also measured in joules. P.E. can be negative.

The work done on a body is equal to its change of mechanical energy which is K.E. + P.E.

4 Conservation of energy

If the weight of the particle is the only force having a component in the direction of motion then throughout the motion:

$$\text{K.E.} + \text{P.E.} = \text{constant}$$

5 Power

Power is the rate of doing work.

For a moving particle:

Power = driving force × speed

Power is measured in watts, where 1 watt (W) is 1 joule per second, or kilowatts, where 1 kilowatt (kW) is 1000 watts.

6 Momentum and Impulse

The momentum of a particle of mass m moving with speed v is mv.

If a constant force $\mathbf{F}$ acts on a particle for time t, then the impulse of the force is equal to $\mathbf{F}t$.

Impulse = change of momentum

For a force in newtons and time in seconds, impulse and momentum are measured in newton-seconds.

7 Conservation of momentum

Momentum is conserved for two particles experiencing a collision or jerk where there are no external forces involved.

$$m_1u_1 + m_2u_2 = m_1v_1 + m_2v_2$$

Moments

6

6.1 Moment of a force

To describe a force acting on a particle you need to know the magnitude and the direction of the force. The effect of the force on the particle is to produce a translation – that is the particle moves in a straight line without turning.

When a force is applied to a rigid body the point of application of the force must be given as well as its magnitude and direction. The force may cause the body to experience a **rotational** motion with or without a translational motion – that is the body may now turn as well as or instead of moving in a straight line.

Suppose you have two forces of unequal magnitude but want them to produce the same **turning effect** about a fixed point. To achieve this, the point of application of the smaller force must be at a greater distance from the fixed point than the point of application of the larger force.

A see-saw can be used to explain this.

A see-saw is usually pivoted at its mid-point and can rotate about this pivot. If two children of unequal weight sit on the see-saw, the heavier child must sit nearer the pivot if the see-saw is to remain horizontal.

The **moment of a force** is a measure of its capability to turn the body on which it is acting.

The moment of the force F about a point P is the product of the magnitude of the force F and the perpendicular distance from the point P to the line of action of F.

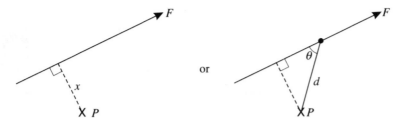

The moment of the force F about P is $F \times x = Fd \sin \theta$.

The moment of the force is measured in newton-metres (Nm) as the force is measured in newtons and the distance in metres.

Sense of rotation

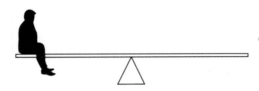

If one child sits on a see-saw on the left-hand side of the pivot, the weight of the child will turn the see-saw in the anticlockwise direction. If the child sits on the right-hand side, the see-saw will turn in a clockwise direction.

This clockwise or anti-clockwise direction of the moment of a force is known as the **sense of rotation**. Use $\circlearrowright$ to show a clockwise direction and $\circlearrowleft$ an anticlockwise direction.

Example 1

A light rod AB is 4 m long and can rotate in a vertical plane about a fixed point C where $AC = 1$ m. A force vertical F of magnitude 8 N acts in a direction perpendicular to AB. Calculate the moment of F about C when F is applied (a) at A (b) at B (c) at C as shown in the diagrams.

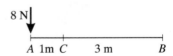

(a) When F is applied at A the moment of F about C is given by:

$$\text{Force} \times \text{perpendicular distance}$$

So:

Moment $\qquad\qquad = 8 \times AC$

Substituting in the value for AC gives:

Moment $\qquad\qquad = 8 \times 1\,\text{Nm}$

$\qquad\qquad\qquad = 8\,\text{Nm anti-clockwise}$

When F is applied at A, the moment is anti-clockwise, because A is on the left-hand side of C and F is acting downwards.

(b) When *F* is applied at *B*:

The moment of *F* about *C* is given by:

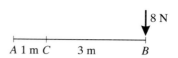

Force × perpendicular distance

So:

Moment   = 8 × CB

Substituting in the value for CB gives:

Moment   = 8 × 3 Nm

                                    = 24 Nm clockwise

When *F* is applied at *B*, the moment is clockwise, because *B* is on the right-hand side of *C* and *F* is acting downwards.

(c) When *F* is applied at *C*:

In this case, the perpendicular distance from C to the force is zero.

So the moment of *F* about *C* is:

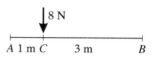

$$8 \times 0 \,\text{Nm} = 0 \,\text{Nm}$$

When *F* is applied at *C*, there is no moment.

Sum of moments

When several forces, all lying in the same plane, act on a body the moments about any point can be added, as long as the sense of rotation for each moment is taken into account.

Example 2

Forces of magnitude 3 N, 4 N and 5 N are applied to a light rod *AB* of length 6 m as shown in the diagram. Calculate the sum of the moments of these forces about *A*.

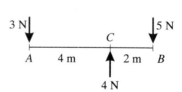

Moment of the 3 N force about *A* is:

$$3 \times 0 \,\text{Nm}$$

Giving 0 Nm.

Moment of the 4 N force about *A* is:

$$4 \times 4 \,\text{Nm}$$

Giving 16 Nm anti-clockwise ↺.
Moment of the 5N force about A is:

$$5 \times 6 \,\text{Nm}$$

Giving 30 Nm clockwise ↻.

As the clockwise moment is larger than the anticlockwise moment, calculate the sum of the moments in the clockwise direction.

So the sum of the moments is:

$$(30 - 16)\,\text{Nm}$$

Giving 14 Nm clockwise ↻.

The sum of the moments about A is 14 Nm clockwise.

Remember that the distance used to calculate a moment must be the *perpendicular* distance from the point to the force. If the length given is not perpendicular to the force, you need to know the angle between that length and the force to calculate the moment of the force.

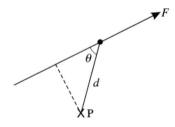

Use moment of the force $= Fd \sin \theta$.

Example 3

The following diagrams show forces acting on a lamina. In each case, calculate the sum of the moments of the forces about the point O.

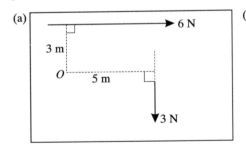

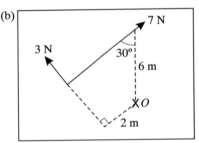

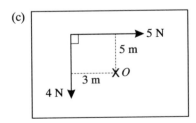

(a) As both forces have a clockwise moment, moment about O is:

$$(6 \times 3 + 3 \times 5) \, \circlearrowright = (18 + 15) \, \circlearrowright$$

$$= 33 \, \text{Nm} \, \circlearrowright$$

The moment about O is 33 Nm, clockwise.

(b) Again both forces have a clockwise moment.

Moment of 7 N force about O is:

$$(7 \times 6 \sin 30) \, \circlearrowright = 21 \, \text{Nm} \, \circlearrowright$$

So the total moment of the forces about O is:

$$(3 \times 2 + 21) \, \circlearrowright = 27 \, \text{Nm} \, \circlearrowright$$

The moment about O is 27 Nm, clockwise.

(c) Here the 5 N force has a clockwise moment and the 4 N force has an anticlockwise moment.

So the moment about O is:

$$(5 \times 5 - 4 \times 3) \, \circlearrowright = 13 \, \text{Nm} \, \circlearrowright$$

The moment about O is 13 Nm clockwise.

Example 4

(In this question **i** and **j** are unit vectors in the directions of the x-and y-axes respectively.)

The force $(6\mathbf{i} + 3\mathbf{j}) \, \text{N}$ is applied at point A of a lamina, where A has position vector $(\mathbf{i} + 4\mathbf{j}) \, \text{m}$ relative to a fixed origin O. Calculate the moment of the force
(a) about the origin O
(b) about the point B with position vector $(2\mathbf{i} - \mathbf{j}) \, \text{m}$.

First draw a clearly labelled sketch.

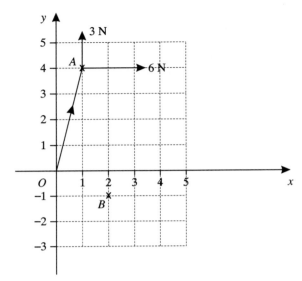

(a) It is convenient to use the components of the force which are parallel to the x-axis (6 N) and the y-axis (3 N), as shown in the diagram.

Use the diagram to read the following distances:

Perpendicular distance from O to 6 N force is 4 m

Perpendicular distance from O to 3 N force is 1 m

Moment of 6 N force is clockwise ⤵

Moment of 3 N force is anticlockwise ⤺

So the total moment about O is:

$$(6 \times 4 - 3 \times 1) \text{⤵} = 21 \text{ Nm ⤵}.$$

The moment about O is 21 Nm clockwise.

(b) Use the diagram to calculate the following distances from B:

Perpendicular distance from B to 6 N force is 5 m

Perpendicular distance from B to 3 N force is 1 m

The moments of both forces are now clockwise

So the total moment about B is:

$$(6 \times 5 + 3 \times 1) \text{⤵} = 33 \text{ Nm ⤵}$$

The moment about B is 33 Nm clockwise.

Exercise 6A

1 Calculate the moment about X of each of the following forces acting on a lamina. State the sense of each moment.

(a)

(b)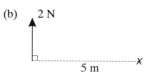

(c)

(d)

(e)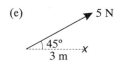

(f)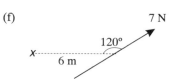

(g)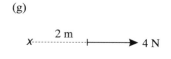

(h)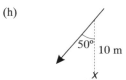

2 The following diagrams show sets of forces acting on a light rod. In each case calculate the sum of the moments about X stating the sense of each sum.

(a)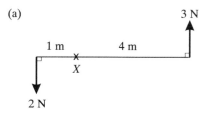

(b)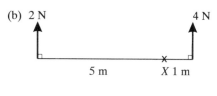

(c)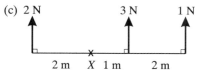

(d)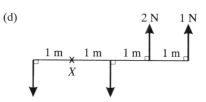

(e)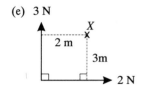

(f)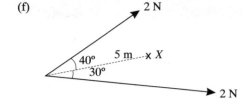

(g)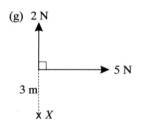

(h)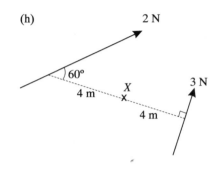

In questions 3 to 8, **i** and **j** are unit vectors in the directions of the *x*- and *y*-axes respectively.

3 The force 5**j** N is applied at point *A* of a lamina where *A* has position vector $(2\mathbf{i} + 4\mathbf{j})$ m relative to a fixed origin. Calculate the moment of the force about the origin.

4 The force $(2\mathbf{i} + 4\mathbf{j})$ N acts at point *P* of a lamina where *P* has position vector 5**j** m relative to a fixed origin. Calculate the moment of the force about the origin.

5 The force 2**j** N acts at the point *A* of a lamina where *A* has position vector 6**i** m relative to a fixed origin. Calculate the moment of the force about the point *B* which has position vector 4**i** m.

6 The force 4**i** N acts at the point *A* of a lamina where *A* has position vector $(2\mathbf{i} + 3\mathbf{j})$ m relative to a fixed origin. Calculate the moment of the force (a) about the origin (b) about the point with position vector $(4\mathbf{i} + 2\mathbf{j})$ m.

7 The force $(5\mathbf{i} + 3\mathbf{j})$ N acts at the point *P* of a lamina where *P* has position vector $(2\mathbf{i} - 2\mathbf{j})$ m relative to a fixed origin. Calculate the moment of the force (a) about the origin (b) about the point which has position vector $(-3\mathbf{i} + 2\mathbf{j})$ m.

8 The force $(3\mathbf{i} - 2\mathbf{j})$ N acts at point *A* of a lamina where *A* has position vector $(2\mathbf{i} + 3\mathbf{j})$ m relative to a fixed origin and the force $(4\mathbf{i} + 3\mathbf{j})$ N acts at the point B of the lamina where B has position vector $(3\mathbf{i} - 2\mathbf{j})$ m. Calculate the sum of the moments of these forces (a) about the origin (b) about the point with position vector $(4\mathbf{i} - 3\mathbf{j})$ m.

6.2 Equilibrium of a lamina under parallel forces

If a lamina is in equilibrium under the action of a system of forces the lamina will not turn about any point. This means the sum of the moments of the forces about any point must be zero. Account must be taken of the sense of the moments when calculating their sum. In chapter 4 it was shown that for a particle to be in equilibrium the resultant force must be zero.

This is also true for a lamina.

- **For a lamina in equilibrium under parallel forces:**
  (a) **the component of the resultant force in any direction must be zero**
  (b) **the algebraic sum of the moments about any point must be zero.**

It follows from (b) that the sum of the anticlockwise moments must equal the sum of the clockwise moments.

One of the forces acting on a lamina which is not light is its weight. The weight acts at the point called the **centre of mass** of the lamina. If the mass of the lamina is evenly distributed, the lamina is said to be **uniform**. The centre of mass of a uniform rod is at the mid-point of the rod. Methods of calculating the position of the centre of mass for other laminae will be discussed later in section 6.3, page 191.

Example 5
A uniform rod AB of length 4 m and mass 5 kg is pivoted at C where $AC = 1$ m. Calculate the mass of the particle which must be attached at A to maintain equilibrium with the rod horizontal.

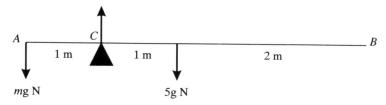

Let the mass of the particle at A be m kg.

The rod is uniform, so the weight of the rod acts at the mid-point of AB.

The pivot at C will exert a vertical force on the rod. The magnitude of this force is not known or required but the moment of this force about C is zero.

Taking moments about C gives:

$$\text{Clockwise moment about } C = 5\,\text{g} \times 1\,\text{Nm} \;\circlearrowright$$

$$\text{Anticlockwise moment about } C = m\text{g} \times 1\text{Nm} \;\circlearrowleft$$

For equilibrium:

$$\text{clockwise moment} = \text{anticlockwise moment}$$

So to calculate the mass of the particle required for equilibrium:

$$5\,\text{g} \times 1 = m\text{g} \times 1$$

$$m = 5$$

The required mass is 5 kg. This means that the mass to be attached at A must be 5 kg to keep the rod in equilibrium.

Example 6

Two children are sitting on a non-uniform plank AB of mass 24 kg and length 2 m. This plank is pivoted at M, the mid point of AB. The centre of mass of AB is at C where AC is 0.8 m. Anne has mass 24 kg and sits at A. John has mass 30 kg.
Find where John must sit for the plank to be horizontal.

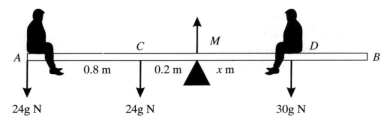

As in example 5, if you take moments about M, the unknown force at the pivot is not required.

Taking moments about M and equating the anticlockwise moments and clockwise moments gives:

$$24\,\text{g} \times 28.8 + 24\,\text{g} \times 0.2 = 30\,\text{g} \times x$$

$$24 + 4.8 = 30x$$

$$x = \tfrac{28.8}{30} = 0.96$$

John should sit 0.96 m from M (or 0.04 m from B).

Example 7

A uniform beam AB of length $5\,\mathrm{m}$ and mass $30\,\mathrm{kg}$ rests horizontally on supports at C and D where $AC = BD = 1\,\mathrm{m}$.

A man of mass $75\,\mathrm{kg}$ stands on the beam at E where $AE = 2\,\mathrm{m}$. Calculate the magnitude of the reaction at each of the supports C and D.

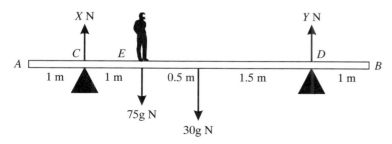

Let the reaction at C be $X\,\mathrm{N}$ and that at D be $Y\,\mathrm{N}$.

To find the two unknown quantities X and Y two equations are needed:

 (i) one obtained by taking moments about some convenient point

(ii) and the other obtained by resolving the forces vertically.

Taking moments about D, abbreviated M(D), will give an equation which involves only one unknown quantity, X.

$$\text{M(D):} \qquad X \times 3 = 75\,g \times 2 + 30\,g \times 1.5$$

$$3X = 150\,g + 45\,g$$

$$X = \frac{195\,g}{3} = \frac{195 \times 9.8}{3}$$

$$= 637$$

The reaction at C has a magnitude of $637\,\mathrm{N}$.

Resolving the forces vertically gives:

$$X + Y = 75\,g + 30\,g$$

Substituting for X gives:

$$Y = 105\,g - 637$$

$$= 392$$

The reaction at D has a magnitude of $392\,\mathrm{N}$.

Alternatively, Y could be calculated by taking moments about C.

M(C): $\qquad \curvearrowleft Y \times 3 = 75\,g \times 1 + 30\,g \times 1.5 \curvearrowright$

$$3Y = 75\,g + 45\,g$$

$$Y = \frac{120\,g}{3} = 392$$

The reaction at D has a magnitude of 392 N.

Exercise 6B

1 The following diagrams show a light rod in equilibrium in a horizontal position. In each case find the magnitudes of the forces X and Y and the distance dm as applicable.

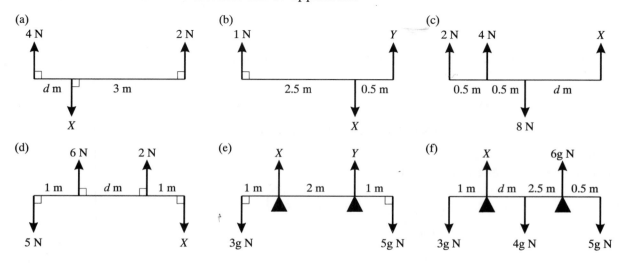

2 A uniform rod AB of length 5 m and mass 6 kg is pivoted at C where $AC = 1.5$ m. Calculate the mass of the particle which must be attached at A to maintain equilibrium with the rod horizontal.

3 A uniform beam AB of length 4 m and mass 21 kg has a particle of mass 15 kg attached at A and a particle of mass 18 kg attached at B. The beam is balanced horizontally on a single support at C. Find the distance AC.

4 A non-uniform rod AB of length 3 m and mass 4 kg rests horizontally on two supports at C and D where $AC = 0.5$ m and $BD = 1$ m. Given that the centre of mass of the rod is at M where $AM = 1.75$ m calculate the magnitudes of the reactions at the supports.

5   A rod *AB* of mass 6 kg and length 3 m has its centre of mass
    1.2 m from *A*. It is suspended horizontally from the ceiling by
    two vertical strings one attached at *A* and the other at the
    point C where *AC* is 2.5 m. Find the magnitude of the
    tensions in the strings.

6   A non-uniform plank *AB* of length 5 m and mass 30 kg is
    pivoted at its mid-point. A child of mass 25 kg sits at *A* and a
    child of mass 35 kg sits at *B*. By modelling the plank as a rod
    and the children as particles find the distance of the centre of
    mass of the plank from *A*.

7   A footbridge across a stream is constructed by placing a tree
    trunk *AB* of length 8 m and mass 90 kg on to supports at *A*
    and *B* so that the tree trunk is horizontal.

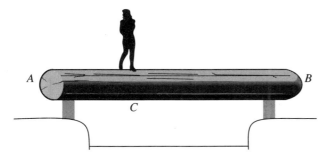

A woman of mass 60 kg stands on the trunk at *C*. The
magnitude of the reaction at *A* is twice the magnitude of the
reaction at *B*. By modelling the tree trunk as a uniform rod
and the woman as a particle calculate (a) the magnitude of
the reaction at *A* and the magnitude of the reaction at *B* (b)
the distance *AC*. Describe the principal difference between
your model and the real-life situation. How could your model
be improved, using further information if necessary?

6.3 Centre of mass of a discrete mass distribution

Every particle in a discrete mass distribution has a force pulling the
particle towards the centre of the earth. This force is called the
weight of the particle. Since the centre of the earth is at a great
distance from the earth's surface, it can be assumed that all these

weights are pulling vertically downwards and that these forces are parallel to each other. The weight of the system is the resultant of the weights of the particles and this resultant force acts through a point called the **centre of mass** of the system.

Centre of mass of a system of particles distributed in one dimension

Consider a simple system consisting of two particles A and B of equal mass m kg. The centre of mass is at the mid-point of the line joining A and B.

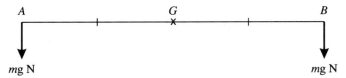

G is the centre of mass of A and B.

Where the particles have different masses or there are more than two particles involved, the method of calculating the centre of mass of the system is shown by example 8.

Example 8

Three particles of masses 5 kg, 3 kg and 4 kg are attached to a light rod AB of length 2 m at the points A, B and C where $AC = 0.6$ m. Find the position of the centre of mass of the system.

Assume the rod to be horizontal and start by drawing two diagrams, one showing the weights of the separate particles and the other showing the resultant weight.

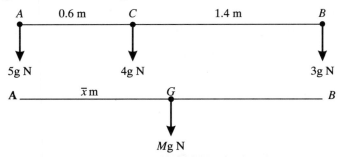

Let the total mass be M kg and the centre of mass of the system be at the point G where $AG = \bar{x}$ m. The forces in the two diagrams are equivalent. So resolving vertically will give the same result for each diagram.

Resolving vertically gives:

$$Mg = 5\,g + 4\,g + 3\,g$$

So : $M = 12$

The total mass is $12\,kg$.

Similarly, both force systems will have the same total moment about any point.

Taking moments about point A gives:

M(A): $\curvearrowright Mg\,\bar{x} = 4\,g \times 0.6 + 3\,g \times 2$

Substituting $M = 12$ gives:

$$12\,g\,\bar{x} = 2.4\,g + 6\,g$$

$$\bar{x} = \frac{8.4}{12} = 0.7$$

The centre of mass of the particles is $0.7\,m$ from A.

This method can be applied to a larger system which has n particles of masses $m_1, m_2, \ldots m_n$ arranged in a straight line where the particles $m_1, m_2, \ldots m_n$ are at a distance $x_1, x_2, \ldots x_n$ respectively from an origin O on the line.

Let the centre of mass of the system be a distance $\bar{x}$ from O.

$$O \quad \underset{\displaystyle \bullet}{\overset{\displaystyle M}{\rule{7cm}{0.4pt}}}$$

In example 8, the weights of the particles were used in the moments equation. Because g was a factor of every term it could be cancelled from the equation. You can therefore work with the masses of the particles rather than the weights.

Then the total mass M for the above system is:

$$M = m_1 + m_2 + m_3 + \ldots + m_n$$

Or: $M = \Sigma m_i$

(Σ is the Greek letter 'sigma' and used here means 'the sum of'.)

Taking moments about O for both diagrams gives:

$$M\bar{x} = m_1 x_1 + m_2 x_2 + m_3 x_3 + \ldots + m_n x_n$$

Or: $M\bar{x} = \Sigma m_i x_i$

Since $M = \Sigma m_i$: $\bar{x} = \dfrac{\Sigma m_i x_i}{\Sigma m_i}$

Centre of mass of a system of particles distributed in two dimensions

If the particles in a system are not in a straight line you need to give their positions as coordinates relative to some fixed axes. The coordinates of the centre of mass can then be found by using the same method as for a one-dimensional distribution (example 8). This time the method must be applied twice, once using x coordinates to give:

$$\bar{x} = \frac{\Sigma m_i x_i}{\Sigma m_i}$$

and again using y coordinates to give:

$$\bar{y} = \frac{\Sigma m_i y_i}{\Sigma m_i}$$

where $(\bar{x}, \bar{y})$ are the coordinates of the centre of mass of the system.

In some questions axes and coordinates are specified, in others you need to choose your own axes.

Using a tabular form to display both the masses of the particles and their coordinates will help you form correct equations.

Example 9

Calculate the coordinates of the centre of mass of particles of mass 4 kg, 3 kg and 2 kg at points with coordinates (2,4), (0,3) and (5,1) relative to axes Ox and Oy.

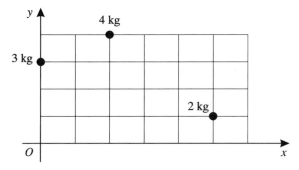

The first step is to find the total mass of the system.

The total mass of the system is $(4 + 3 + 2)\,\text{kg} = 9\,\text{kg}$.

Now form a table using the masses of the particles and their coordinates. The final column will contain the total mass and the unknown coordinates $(\bar{x}, \bar{y})$ of the centre of mass. Separate this column from the others with a double line.

	Separate masses			Total mass
Mass (kg)	3	4	2	9
x coordinate	0	2	5	$\bar{x}$
y coordinate	3	4	1	$\bar{y}$

Using the x coordinates gives:

$$\Sigma m_i x = \Sigma m_i \bar{x}$$
$$3 \times 0 + 4 \times 2 + 2 \times 5 = 9\bar{x}$$
$$9\bar{x} = 18$$
$$\bar{x} = 2$$

Using the y coordinates gives:

$$\Sigma m_i y_i = \Sigma m_i \bar{y}$$
$$3 \times 3 + 4 \times 4 + 2 \times 1 = 9\bar{y}$$
$$9\bar{y} = 27$$
$$\bar{y} = 3$$

The coordinates of the centre of mass are (2,3).

Example 10

Particles of masses 1 kg, 2 kg, 3 kg and 4 kg are attached to the corners of a light rectangular plate $ABCD$. Given that $AB = 10$ cm and $AD = 5$ cm calculate the distance of the centre of mass of the system (a) from AB (b) from AD.
In this question no axes are given, but you need to find the distances of the centre of mass from AB and AD.

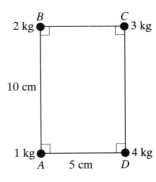

These lines are perpendicular to each other and are the best choice of axes.

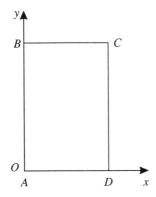

The total mass of the system is $(1 + 2 + 3 + 4)$ kg $= 10$ kg.

Let the centre of mass be at G where G is $\bar{x}$ cm from AB and $\bar{y}$ cm from AD.

	Separate masses				Total mass
Mass (kg)	1	2	3	4	10
Distance from AB (cm)	0	0	5	5	$\bar{x}$
Distance from AD (cm)	0	10	10	0	$\bar{y}$

Using the x coordinates gives:

$$\Sigma m_i x_i = \Sigma m_i \bar{x}$$
$$1 \times 0 + 2 \times 0 + 3 \times 5 + 4 \times 5 = 10\bar{x}$$
$$10\bar{x} = 35$$
$$\bar{x} = 3.5$$

The centre of mass is 3.5 cm from AB.

Using the y coordinates gives:

$$\Sigma m_i y_i = \Sigma m_i y$$
$$1 \times 0 + 2 \times 10 + 3 \times 10 + 4 \times 0 = 10\bar{y}$$
$$10\bar{y} = 50$$
$$\bar{y} = 5$$

The centre of mass is 5 cm from AD.

The above method can also be used if the position of the centre of mass of the system is given and either the position or the mass of one of the particles needs to be determined.

Example 11

Particles A, B and C of mass 2 kg, 4 kg and 4 kg are situated at points with coordinates (3,0), (1,3) and (0,1) respectively. A fourth particle D of mass 2 kg is placed at the point with co-ordinates (x, y). Given that the centre of mass of the resulting system is at (1,2) determine the values of x and y.

	Separate masses				Total mass
Mass (kg)	2	4	4	2	12
x coordinate	3	1	0	x	1
y coordinate	0	3	1	y	2

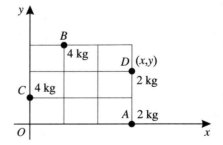

Using the x coordinates gives:

$$\Sigma m_i x = \Sigma m_i \bar{x}$$
$$2 \times 3 + 4 \times 1 + 4 \times 0 + 2x = 12 \times 1$$
$$2x = 2$$
$$x = 1$$

Using the y coordinates gives:

$$\Sigma m_i y_i = \Sigma m_i \bar{y}$$
$$2 \times 0 + 4 \times 3 + 4 \times 1 + 2y = 12 \times 2$$
$$2y = 8$$
$$y = 4$$

So x is 1 and y is 4.

If any of the particles are situated at points which have negative values for either or both coordinates the method used is the same but the signs of the coordinates must be included.

Example 12

Particles A, B, C and D of masses 1 kg, 2 kg, 1 kg and m kg respectively are situated at points with coordinates $(5,-1)$, $(-1,2)$, $(6,4)$ and $(1,0)$ respectively. Given that the centre of mass of the system is at $(2, \bar{y})$, determine the values of m and $\bar{y}$.

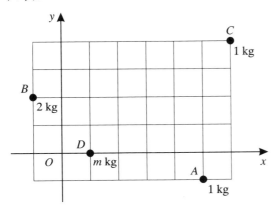

	Separate masses				Total mass
Mass (kg)	1	2	1	m	$4 + m$
x coordinate	5	-1	6	1	2
y coordinate	-1	2	4	0	$\bar{y}$

Using the x coordinates gives:

$$\Sigma m_i x_i = \Sigma m_i \bar{x}$$
$$1 \times 5 + 2 \times (-1) + 1 \times 6 + m \times 1 = (4 + m)2$$
$$5 - 2 + 6 + m = 8 + 2m$$
$$1 = m$$

Using the y coordinates gives:

$$\Sigma m_i y_i = \Sigma m_i y$$
$$1 \times -1 + 2 \times 2 + 1 \times 4 + m \times 0 = (4 + m)\bar{y}$$
$$-1 + 4 + 4 = 5\bar{y}$$

Substituting $m = 1$ gives:
$$7 = 5y$$
$$\bar{y} = \tfrac{7}{5} = 1.4$$

So m is 1 and $\bar{y}$ is 1.4.

Exercise 6C

1   Three particles of masses 1 kg, 2 kg and 3 kg lie on the x-axis
    at points with coordinates (1,0), (3,0) and (6,0) respectively.
    Determine the coordinates of the centre of mass of the
    system.

2   Four particles of masses 4 kg, 3 kg, 2 kg and 5 kg lie on the
    y-axis at points with coordinates (0,2), (0,4), (0,5) and (0,6)
    respectively. Determine the coordinates of their centre of
    mass.

3   Three particles of masses 2 kg, 4 kg and 4 kg lie at the points
    with coordinates (1,1), (3,3) and (4,4) respectively. Determine
    the coordinates of their centre of mass.

4   A light rod AB of length 3 m has particles of masses 2 kg, 4 kg
    and 3 kg attached to it at the points A, B and C respectively
    where $AC = 1$ m. Determine the distance of the centre of
    mass from A.

5   Three particles of masses 3 kg, 5 kg and m kg lie on the x-axis
    at the points with coordinates (2,0), (4,0) and (5,0)
    respectively. The centre of mass of the system is at (4,0).
    Determine the value of m.

6   A light rod AB of length $2m$ has particles of masses 4 kg and
    6 kg attached to it at the points A and C respectively where
    $AC = 0.5$ m. Determine the mass of the particle that must be
    attached at B if the centre of mass of the system is to be at
    the mid-point of the rod.

7   Three particles of masses 2 kg, 3 kg and 2 kg lie in the (x, y)
    plane at the points with coordinates (2,3), (3,6), and (−3,2)

respectively. Determine the coordinates of the centre of mass of the system.

8 Four particles of equal mass lie in the (x, y) plane at the points with coordinates $(3,5)$, $(-1,2)$, $(3,-4)$ and $(-3, -2)$. Determine the coordinates of the centre of mass of the system.

9 A light rectangular plate $PQRS$ where $PQ = 4\,$m and $PS = 3\,$m has particles of masses $2\,$kg, $2\,$kg, $3\,$kg and $4\,$kg attached at P, Q, R and S respectively. Calculate the distance of the centre of mass of the system (a) from PQ (b) from PS.

10 A light rectangular plate $ABCD$ where $AB = 10\,$cm and $AD = 9\,$cm has particles of masses $3\,$kg, $2\,$kg and $4\,$kg attached to it at the points A, B and X respectively where X is the mid-point of CD. Calculate the distance of the centre of mass of the system from (a) AB (b) AD.

11 A light rectangular plate $ABCD$ where $AB = 5\,$cm and $AD = 12\,$cm has particles of masses $4\,$kg, $3\,$kg and $2\,$kg attached at A, B and C respectively. A fourth particle of mass $5\,$kg is to be attached to the plate. Given that the centre of mass of the resulting system is to be at the centre of the plate, determine the distance at which this particle must be attached (a) from AB (b) from AD.

12 Particles of masses $4\,$kg, $3\,$kg and $5\,$kg are attached to the corners A, B and C of a light rectangular plate $ABCD$ where $AB = 2\,$m and $AD = 4\,$m. Calculate the mass that must be attached at D if the centre of mass of the system is to be at a distance of $2\,$m from AB. Using this mass, calculate the distance of the centre of mass from AD.

13 A rectangular framework is made by joining light rods AB, BC, CD and AD where $AB = CD = 1\,$m and $BC = AD = 2\,$m. Particles of masses $2\,$kg, $4\,$kg, $4\,$kg and $5\,$kg are placed at the mid-points of AB, BC, CD and AD respectively. Calculate the distances of the centres of mass of the resulting system from (a) AB (b) AD.

6.4 Centre of mass of a uniform plane lamina

A uniform plane lamina has its mass distributed evenly throughout its area. Any uniform plane lamina with an axis of symmetry has its mass evenly distributed on either side of the axis. Therefore the centre of mass of that figure must lie on the axis of symmetry. For a figure with more than one axis of symmetry, it follows that the centre of mass is at the point of intersection of these axes.

■ **The symmetries of a plane lamina can be used to determine its centre of mass.**

Standard results

Uniform rectangular lamina
A uniform rectangular lamina has two axes of symmetry which are the two lines joining the mid-points of opposite sides. The centre of mass G is at their point of intersection.

Uniform circular disc
A uniform circular disc has every diameter as an axis of symmetry. The centre of the circle is therefore the centre of mass G of the disc.

Uniform triangular lamina
A uniform triangular lamina only has axes of symmetry if it is equilateral or isosceles.

A uniform equilateral triangle has three axes of symmetry which are called the **medians** of the triangle. These are the three lines which join the vertices of the triangle to the mid-points of the opposite sides. The centre of mass G is at their point of intersection.

A uniform isosceles triangle has only one axis of symmetry. Its centre of mass still lies at the point of intersection of the medians.

A scalene triangle has no axes of symmetry. It can be shown that the medians of any triangle intersect at a point which is the centre of mass G of the triangle. This point is $\frac{1}{3}$ distance along the median from the mid-point of that side. That is for triangle ABC opposite, $GD = \frac{1}{3} AD$.

This result is explained in Mechanics 2 (Book M2).

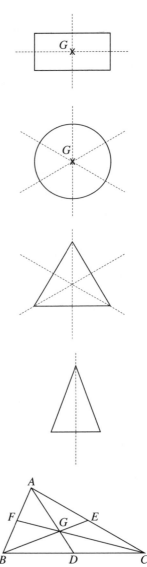

Example 13

Calculate the coordinates of the centre of mass of triangle ABC with vertices at (0,0), (0,9) and (12,0).

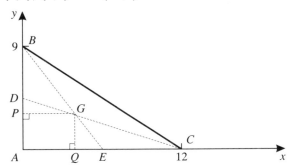

Let D be the mid-point of AB and E be the mid-point of AC.

Then the centre of mass is at G, where $CG = \frac{2}{3} CD$ and $BG = \frac{2}{3} BE$.

Because the triangles ADC and QGC are similar:

$$\frac{CG}{CD} = \frac{CQ}{AC}$$

And because:

$$\frac{CG}{CD} = \frac{2}{3}$$

It follows that:

$$\frac{CQ}{AC} = \frac{2}{3}$$

Or:

$$CQ = \frac{2}{3} AC$$

Then $AQ = \frac{1}{3} AC$ and the x coordinate of G is 4.

Also because the triangle BPG and BAE are similar:

$$\frac{BP}{BA} = \frac{BG}{BE} = \frac{2}{3}$$

So: $BP = \frac{2}{3} AB$ and $AP = \frac{1}{3} AB$.

The y coordinate of G is 3.

The coordinates of G are (4,3).

In questions like the above, where the answer can be 'seen' from the geometry of the diagram it is not necessary to write down a lengthy calculation.

Composite plane figures

Combining two or more uniform plane figures, like those above, will produce a composite figure.

You can calculate the position of the centre of mass of a composite figure by using the same methods as before with the standard results given above.

If the composite figure has an axis of symmetry then the centre of mass will lie on that axis.

Example 14

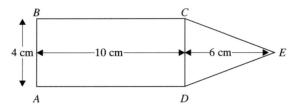

This diagram shows a uniform lamina formed by joining a rectangular lamina $ABCD$ and an isosceles triangular lamina CDE, both of mass per unit area m kg. Calculate the distance of the centre of mass of $ABCED$ (a) from AB (b) from AD.
Take the x- and y-axes to be in the directions of AD and AB respectively.

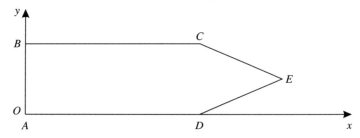

The area of $ABCD$ is:

$$(4 \times 10) = 40 \, \text{cm}^2$$

And the mass of $ABCD = 40 \, m$ kg.
Similarly the mass of CDE is:

$$(\tfrac{1}{2} \times 4 \times 6 \, \text{m}) = 12 \, m \, \text{kg}.$$

So the total mass of $ABCED$ is $(40 \, m + 12 \, m) = 52 \, m$ kg.

Construct a table showing the masses and coordinates of the centres of masses of the separate parts and of the total mass.

The distance x is the distance of the centre of mass from AB and the distance y is the distance of the centre of mass from AD.

	Separate masses		Total mass
Mass (kg)	$40 \, m$	$12 \, m$	$52 \, m$
Distance x (cm)	5	$(10 + 2) = 12$	$\bar{x}$
Distance y (cm)	2	2	$\bar{y}$

(a) Using the x coordinates gives:

$$40\,m \times 5 + 12m \times 12 = 52\,m \times \bar{x}$$

$$\bar{x} = \frac{344}{52} = 6.62$$

The centre mass is 6.62 cm from AB.

(b) Using the y coordinates gives:

$$40m \times 2 + 12m \times 2 = 52\,m\bar{y}$$

$$\bar{y} = \frac{104}{52} = 2$$

The centre of mass is 2 cm from AD.

There is an alternative solution for (b).

If you see an axis of symmetry for the figure you can use this to obtain your solution but you must state that you are using symmetry.

In example 14, the lamina has an axis of symmetry which is the line joining the midpoint of AB to E. The centre of mass must lie on this line. This can be used to solve part (b) of the question.

In this case, using symmetry for (b), the centre of mass is 2 cm from AD.

Because m, the mass per unit area, was a factor for every term in both the equations in example 14, all the masses cancelled out. When calculating the position of the centre of mass of a uniform lamina you only need to use the ratio of the masses as given by the areas.

If you are required to find the position of the centre of mass of a lamina which has had part of its original area cut away, use the same method so long as the complete lamina is considered to be formed from the final lamina and the part cut away.

Example 15
An ear-ring is formed from a thin uniform square of metal $ABCD$ of side 2 cm by cutting away a square $PQRS$ of side 0.6 cm, as shown in the diagram.

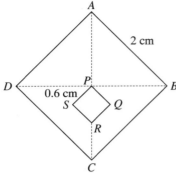

P lies on the diagonal BD of $ABCD$ and PR lies along the diagonal AC of $ABCD$.

Find the position of the centre of mass of the ear-ring.

Diagonal AC is an axis of symmetry of the ear-ring.

So the centre of mass lies on AC.

You need the distances from A of the centres of mass of the two squares $ABCD$ and $PQRS$.

Centre of mass of $ABCD$ is at P.

$$AP = \tfrac{1}{2} AC$$

By Pythagoras' Theorem:

$$AC = \sqrt{2^2 + 2^2} = 8 \text{ cm}$$

Hence: $$AP = \tfrac{\sqrt{8}}{2} = 1.41$$

So the distance of the centre of mass of $ABCD$ from A is $1.41\,$cm.

Centre of mass of $PQRS$ is at the mid-point of PR.

$$PR = \sqrt{(0.6^2 + 0.6^2)}$$
$$= 0.8485$$

So the distance of the centre of mass of $PQRS$ from A is:

$$(1.414 + \tfrac{1}{2} \times 0.8485) \text{ cm}$$
$$= 1.838 \text{ cm}$$

Let the centre of mass of the ear-ring be distance $\bar{x}\,$cm from A, measured along AC.

$$\text{Area of } PQRS = 0.36 \text{ cm}^2$$
$$\text{Area of ear-ring} = 4 - 0.36$$
$$= 3.64 \text{ cm}^2$$

	Separate masses		Total mass
Ratio of masses	0.36	3.64	4
Distance of C of M from A	1.838	$\bar{x}$	1.414

Taking moments gives:

$$0.36 \times 1.838 + 3.64\bar{x} = 4 \times 1.414$$

$$\bar{x} = \frac{4 \times 1.414 - 0.36 \times 1.836}{3.64}$$

$$\bar{x} = 1.37.$$

The centre of mass is on AC at a distance 1.37 cm from A.

Centre of mass of a uniform rod and of a framework

The centre of mass of a uniform rod is at the mid-point of the rod.

A **framework** is a body which is formed by joining a number of rods or wires. The position of the centre of mass of a framework can be calculated by using the positions of the centres of mass of the separate rods or wires.

Example 16

A uniform wire of length 30 cm is bent to form a triangle ABC where AB has length 5 cm and AC has length 12 cm. Calculate the distance of the centre of mass (a) from AB (b) from AC.

The perimeter of the triangle is 30 cm and two sides are 5 cm and 12 cm. So BC is 13 cm and the triangle is right-angled at A.

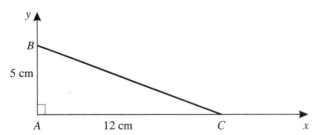

Take axes along AC and AB as shown in the diagram. Since the wire is uniform the mass of each side of the triangle will be proportional to its length. Each side of the triangle is a uniform rod and so the centre of mass of each side is at its mid-point.

Let the centre of mass of the triangle have coordinates $(\bar{x}, \bar{y})$ where $\bar{x}$ is the distance of the centre of mass from AB and $\bar{y}$ is the distance of the centre of mass from AC.

	Separate masses			Total mass
Ratio of masses	5	12	13	30
x coordinate	0	6	6	$\bar{x}$
y coordinate	2.5	0	2.5	$\bar{y}$

Using the x coordinates gives:

$$5 \times 0 + 12 \times 6 + 13 \times 6 = 30\bar{x}$$

$$72 + 78 = 30\bar{x}$$

$$\bar{x} = \frac{150}{30} = 5$$

The centre of mass is 5 cm from AB.

Using the y coordinates gives:

$$5 \times 2.5 + 12 \times 0 + 13 \times 2.5 = 30\bar{y}$$

$$\bar{y} = \frac{45}{30} = 1.5$$

The centre of mass is 1.5 cm from AC.

Exercise 6D

1 For each of the following uniform laminae write down the coordinates of the centre of mass :

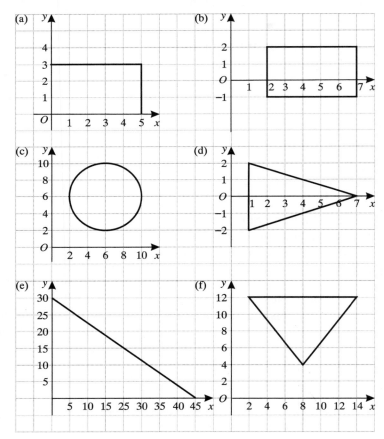

2 For each of the following uniform laminae calculate the coordinates of the centre of mass.

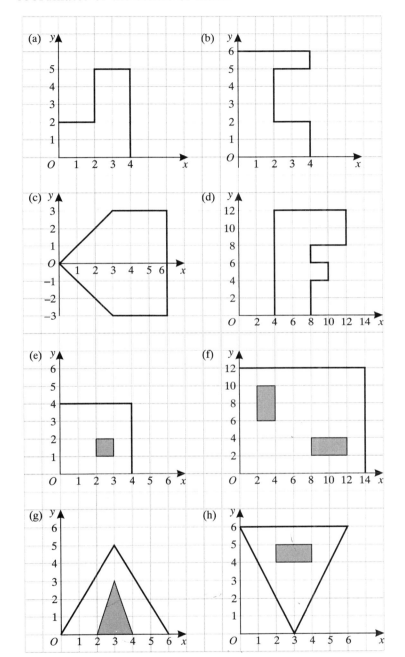

3 A uniform wire of length 6 m is bent to form a triangle ABC where AB is 1.5 m and BC is 2.5 m. Calculate the distance of the centre of mass of the triangle (a) from AB (b) from AC.

4 Three uniform rods, all of the same mass per unit length, are joined to form a triangle ABC where AB is 7 cm, BC is 24 cm and AC is 25 cm. Calculate the distance of the centre of mass of the triangle (a) from AB (b) from BC.

5 Three uniform rods AB, BC and AC are of length 3 m, 4 m and 5 m and have masses 5 kg, 4 kg and 3 kg respectively. They are joined to make a framework. Calculate the distance of the centre of mass of the framework (a) from AB (b) from BC.

6 A uniform circular lamina of radius 5 units has its centre at (2,0). Two circles are cut away from this lamina, one having centre (0,0) and radius 2 units and the other having centre (3,2) and radius 1 unit. Calculate the coordinates of the centre of mass of the resulting lamina.

7 A uniform square metal plate of side 10 cm has a square of side 4 cm cut away from one corner. Find the position of the centre of mass of the remaining plate.

8 The diagram shows a thin uniform square of metal which has a corner folded over. Calculate the distance of the centre of mass of $ABCEF$ from the corner A.

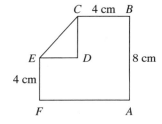

6.5 Equilibrium of a plane lamina

A lamina can be in equilibrium in a vertical plane when it is suspended from a point or is balanced on an edge. For either of these situations you need to know the position of the centre of mass of the lamina before you can proceed but the situations themselves are very different and so they will be considered separately.

Suspension of the lamina from a fixed point

If a lamina is suspended by a string and allowed to hang freely there are only two forces acting on it – its weight and the tension in the string. Consider a triangle ABC suspended from vertex A.

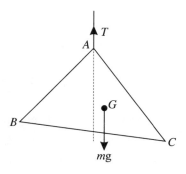

ABC could not hang as shown in the diagram because its weight would have a clockwise moment about A and the tension would have a zero moment about the same point. Hence the total moment cannot be zero as needed for equilibrium. The lamina must be in the position where the line of action of the weight passes through A so that there is no resultant moment about A.

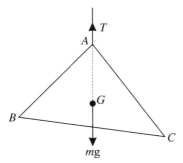

■ **For a suspended lamina to be in equilibrium its centre of mass, G, must be vertically below the point of suspension, A.**

Example 17

The lamina shown in the diagram is suspended freely from A. Find the angle between AB and the vertical.

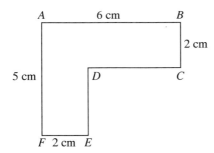

First find the position of the lamina's centre of mass, G. Consider the lamina to be formed of two rectangles $ABCP$ and $PDEF$.

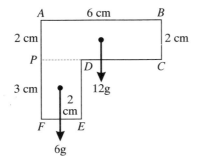

Let the centre of mass have coordinates $(\bar{x}, \bar{y})$ where $\bar{x}$ is the distance of the centre of mass from AF and $\bar{y}$ is the distance from AB.

	Separate masses		Total mass
Ratio of masses	12	6	18
Distance x (cm)	3	1	$\bar{x}$
Distance y (cm)	1	3.5	$\bar{y}$

Using the x coordinates gives:

$$12 \times 3 + 6 \times 1 = 18\bar{x}$$

$$\bar{x} = \frac{42}{18} = 2.333$$

Using the y coordinates gives:

$$12 \times 1 + 6 \times 3.5 = 18\bar{y}$$

$$\bar{y} = \frac{33}{18} = 1.833$$

To find the angle between AB and the vertical when the lamina is suspended, it is not necessary to re-draw the lamina. Simply add G to your original diagram, join AG and label this line as being vertical.

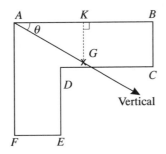

Mark the angle required as θ on your diagram and add the perpendicular line GK from G to AB to make a right-angled triangle. The lengths of AK and GK are the distances of G from AF and AB respectively.

So:

$$\tan \theta = \frac{GK}{AK} = \frac{1.833}{2.333}$$

$$\theta = 38.16°$$

AB is inclined at $38.2°$ to the vertical.

Equilibrium of a lamina on an inclined plane

If a lamina is balanced on an inclined plane, the line of action of the weight of the lamina falls within the side of the lamina in contact with the plane. The weight will have a clockwise moment about the point A but the plane prevents the lamina from turning.

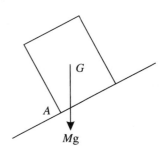

The angle of the plane could be increased, so that the situation is as shown opposite.

The weight now has an anticlockwise moment about A and the plane is no longer preventing the lamina from turning. In this case equilibrium is not possible.

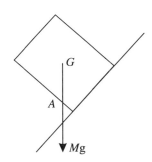

■ **For a lamina on an inclined plane to be in equilibrium the line of action of the weight of the lamina must fall within the side of the lamina which is in contact with the plane.**

Example 18

$ABCD$ is a rectangular lamina with $AB = 12$ cm and $AD = 5$ cm. The lamina rests in equilibrium on an inclined plane with AD in contact with the plane. Find the maximum possible angle of inclination of the plane.

Let the plane be inclined at angle θ.

Let G be the centre of mass of the lamina.

θ will be maximum when AG is vertical, since increasing the angle of the plane any more will cause the line of action of the weight to fall outside the line of contact between the lamina and the plane.

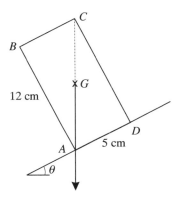

From the geometry of the diagram $\angle BAG = \theta$

G is at the mid-point of the diagonals of $ABCD$.

So A, G and C are collinear.

Therefore:
$$\tan \angle BAG = \tfrac{5}{12}$$
$$\angle BAG = 22.6°$$

So:
$$\theta = 22.6°$$

The maximum angle of inclination of the plane is 22.6°.

Exercise 6E

1 A uniform rectangular lamina $ABCD$ has $AB = 5$ cm and $AD = 8$ cm. It is suspended from A and hangs freely under gravity. Calculate the angle between AB and the downwards vertical.

2 A uniform triangular lamina ABC where $AB = 1$ m and $BC = 2$ m is right-angled at B. It is suspended from C and hangs freely under gravity. Calculate the angle between BC and the downwards vertical.

3 The framework described in Question 5 of Exercise 6D is suspended from B and hangs freely under gravity. Calculate the angle between AB and the vertical.

4 The wire described in Question 3 of Exercise 6D is hanging freely under gravity. Calculate the angle between AC and the vertical (a) when the wire is suspended from A (b) when the wire is suspended from C.

5 The ear-ring shown in the diagram is made from a uniform sheet of plastic. It consists of a semicircle centre O of radius 2 cm from which a concentric semicircle of radius 1 cm has been cut away. Given that the centre of mass of a uniform semicircular lamina of radius r is on the axis of symmetry at a distance $\dfrac{4r}{3\pi}$ from the straight edge, model the ear-ring as a uniform lamina and calculate the distance of its centre of mass from O. The ear-ring is suspended from point A and hangs freely under gravity. Calculate the angle between AB and the vertical, giving your answer to the nearest degree.

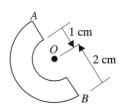

6 An engineer's square is in the shape of two uniform rectangular pieces of metal joined as shown in the diagram. The mass of $BCDE$ is three times the mass of $ABFG$. By modelling each rectangle as a uniform lamina calculate the distance of the centre of mass of the square (a) from AC (b) from CD.

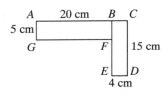

The square is placed on a peg at F. By modelling the square as a lamina freely suspended from F, calculate the angle between FE and the vertical when the square hangs in equilibrium. State two assumptions you have made when using this model to calculate the angle.

7 A uniform rectangular lamina $ABCD$ where $AB = 10$ cm and $BC = 5$ cm is placed on an inclined plane as shown in the diagram.

(a) Determine whether the rectangle can remain in equilibrium when the plane is inclined at an angle of $25°$.

(b) Calculate the maximum possible angle of inclination of the plane if the rectangle is to be balanced as shown.

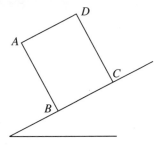

8 The metal plate described in Question 7 of Exercise 6D is placed on an inclined plane as shown in the diagram.

The plane is inclined to the horizontal at an angle θ.
Determine whether equilibrium is possible (a) when $\theta = 10°$
(b) when $\theta = 25°$.
Calculate the maximum possible angle of inclination of the
plane for the lamina to be in equilibrium.

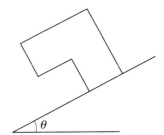

9   A uniform triangular sheet of metal ABC where $AB = 6\,\text{cm}$
and $AC = 12\,\text{cm}$ is right-angled at A. The triangle is placed
on a plane inclined to the horizontal at angle θ as shown in
the diagram.
By modelling the triangle as a uniform lamina determine the
maximum value of θ for the triangle to remain in equilibrium
in this position. As the angle of inclination of the plane is
increased from zero, the equilibrium could be broken in a
different way from tilting. State the form of motion that
could have taken place. State also the assumption you made
about your original model so that this form of motion did
not take place.

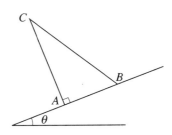

10 The engineer's square described in Question 6 of this exercise
is placed on a plane inclined at angle θ to the horizontal as
shown in the diagram.
Find the maximum value of θ if the square is to remain in
equilibrium.
The square is now placed on the inclined plane as shown in
the second diagram.
Determine whether equilibrium is possible (a) when $\theta = 10°$
(b) when $\theta = 30°$.

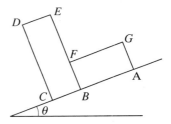

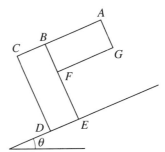

SUMMARY OF KEY POINTS

1 The moment of a force of magnitude F about a point P is given by:

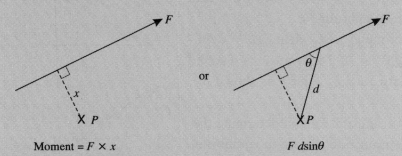

Moment $= F \times x$ or $F\,d\sin\theta$

For a force in newtons and a distance in metres the moment is measured in newton-metres (Nm).

Moments can be clockwise or anticlockwise in sense.

2 For a lamina to be in equilibrium under parallel forces:

(a) the resultant force in any direction must be zero

(b) the component of the resultant force in any direction must be zero

(c) the algebraic sum of the moments about any point must be zero.

3 The centre of mass of a lamina is the point at which the weight acts.

The weight of a uniform lamina is evenly distributed. The centre of mass for a lamina or discrete mass distribution must lie on any axis of symmetry.

The centre of mass of a set of n masses $m_1, m_2, \ldots m_n$ at points with coordinates (x_1, y_1), (x_2, y_2), $\ldots (x_n, y_n)$ has coordinates $(\bar{x}, \bar{y})$ where:

$$\bar{x} = \frac{\Sigma m_i x_i}{\Sigma m_i}$$

And: $$\bar{y} = \frac{\Sigma m_i y_i}{\Sigma m_i}$$

4 Standard results:

Body	Centre of mass
Uniform rod	mid-point of rod
Uniform rectangular lamina	point of intersection of lines joining mid-points of opposite sides
Uniform circular disc	centre of circle
Uniform triangular lamina	point of intersection of medians – in other words $\frac{2}{3}$ distance from any vertex to the mid-point of the opposite side

5 For a lamina which is freely suspended and hangs in equilibrium, the centre of mass will be vertically below the point of suspension.

For a lamina which is balanced on an inclined plane, the line of action of the weight must fall within the side of the lamina that is in contact with the plane.

Review exercise

Whenever a numerical value of g is required take g = 9.8 m s^{-2}.

1   A book of mass 2 kg rests on a rough plane inclined at an angle $\alpha°$ to the horizontal. Given that the coefficient of friction between the book and the plane is 0.2, and that the book is on the point of slipping down the plane, find, to the nearest degree, the value of α.   [L]

2   A block of mass 3 kg rests on a rough, horizontal table. When a force of magnitude 10 N acts on the block at an angle of 60° to the horizontal in an upward direction, the block is on the point of slipping. Calculate, to 2 significant figures, the value of the coefficient of friction between the block and the table.   [L]

3   A particle P of mass 7m is placed on a rough horizontal table, the coefficient of friction between P and the table being μ. A force of magnitude 2mg, acting upwards at an acute angle α to the horizontal, is applied to P and equilibrium is on the point of being broken by the particle sliding on the table. Given that $\tan \alpha = \frac{5}{12}$, find the value of μ.   [L]

4   (a) A book is placed on a desk lid which is slowly tilted. Given that the book begins to slide when the inclination of the lid to the horizontal is 30°, find the coefficient of friction between the book and the desk lid.   [L]

(b) State an assumption you have made about the book when forming the mathematical model you used to solve part (a).

5   A particle is placed on a smooth plane inclined at 35° to the horizontal. The particle is kept in equilibrium by a horizontal force, of magnitude 8 N, acting in the vertical plane containing the line of greatest slope of the inclined plane through the particle. Calculate, in N to one decimal place,

(a) the weight of the particle

(b) the magnitude of the force exerted by the plane on the particle. [L]

6 A particle is suspended by two light inextensible strings and hangs in equilibrium. One string is inclined at 30° to the horizontal and the tension in that string is of magnitude 40 N. The second string is inclined at 60° to the horizontal. Calculate in N (a) the weight of the particle (b) the magnitude of the tension in the second string. [L]

7 A rough plane is inclined at an angle α to the horizontal, where $\tan \alpha = \frac{3}{4}$. A particle slides with acceleration $3.5 \, \text{m s}^{-2}$ down a line of greatest slope of this inclined plane. Calculate the coefficient of friction between the particle and the inclined plane. [L]

8 A particle starts from rest and slides with acceleration $3 \, \text{m s}^{-2}$ down a fixed plane which is inclined at 40° to the horizontal. Calculate, to 2 significant figures, the coefficient of friction between the particle and the plane. [L]

9 (a) Forces $(3\mathbf{i} + a\mathbf{j})$ newtons and $(b\mathbf{i} - 4\mathbf{j})$ newtons act on a body of mass 2 kg. The acceleration of the body is $(5\mathbf{i} - 3\mathbf{j}) \, \text{m s}^{-2}$. Find the values of a and b. [L]

(b) Part (a) refers to 'the body'. What assumption did you make about 'the body' when solving the question?

10 A particle P of mass 2 kg is acted on by a force of $(10\mathbf{i} - 24\mathbf{j}) \, \text{N}$.

(a) Find, in m s^{-2}, the acceleration vector of P and the magnitude of this acceleration.

(b) Given that P starts from rest find, in m, the distance travelled by P in the first 3 seconds of the motion. [L]

11 A hot-air balloon, its occupants and ballast have total mass 2000 kg. The balloon is travelling horizontally with constant velocity. The occupants release 100 kg of ballast. Neglecting air resistance find, in m s^{-2} to 2 significant figures, the magnitude of the acceleration of the balloon. [L]

12 A particle of mass 4 kg rests on a smooth horizontal plane. Forces of 8 N due east and 11 N due south are applied to the particle. Calculate the acceleration produced, giving its magnitude in m s^{-2} to 2 significant figures and its direction as a bearing to the nearest degree.   [L]

13 A block of mass 3 kg is pulled along a rough horizontal floor by a constant force of magnitude 20 N inclined at an angle of 60° to the upward vertical. The acceleration of the block has magnitude 2 m s^{-2}. Calculate, to 2 decimal places, the value of the coefficient of friction between the block and the floor. [L]

14 At a particular instant, the engine of a motor launch is producing a driving force of magnitude 6000 N and the launch is accelerating at 2 m s^{-2}. Given that the mass of the launch is 2300 kg find, in N, the magnitude of the force opposing the motion of the launch.   [L]

15

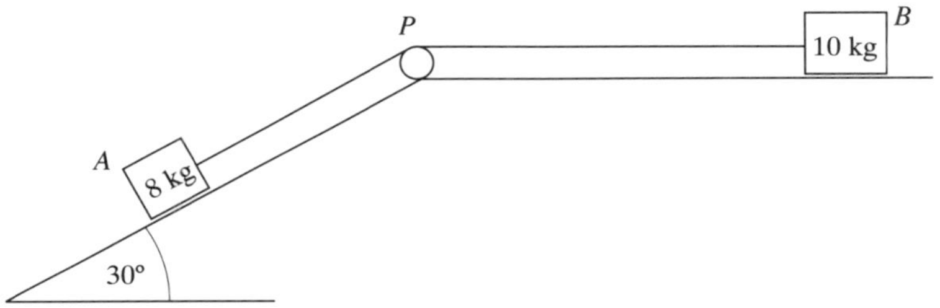

Two particles A and B, of mass 8 kg and 10 kg respectively, are connected by a light inextensible string which passes over a light smooth pulley P. Particle B rests on a smooth horizontal table and particle A rests on a smooth plane inclined at 30° to the horizontal with the string taut and perpendicular to the line of intersection of the table and the plane as shown above. The system is released from rest.

(a) Find the magnitude of the acceleration of particle B.

(b) Find, in newtons, the tension in the string.

(c) Find the distance covered by B in the first two seconds of motion, given B does not reach the pulley.   [L]

16 (a) A tug, of mass 9000 kg, is pulling a boat, of mass 5000 kg, by means of an inextensible horizontal tow rope along a straight canal. The resistive forces opposing the motions of the tug and the boat are 1500 N and 800 N respectively. Calculate, in N, the tension in the rope when the tug is accelerating at $\frac{1}{2}$ m s^{-2}.   [L]

(b) How did you model the tug and boat to answer part (a)?

(c) The tow rope is described as 'inextensible'. Do you think this is possible in real-life? Briefly justify your answer.

17 A particle A, of mass 0.8 kg, resting on a smooth horizontal table 1 m from the edge, is connected to a particle B, of mass 0.6 kg, which is 1 m from the ground, by a light inextensible string passing over a small smooth pulley fixed at the edge of the table. The system is released from rest with the horizontal part of the string perpendicular to the edge of the table, the hanging part vertical and the string taut. Calculate:

(a) the acceleration, in m s^{-2}, of A

(b) the tension, in N, in the string

(c) the speed, in m s^{-1}, of B when it hits the ground

(d) the time, in s, taken for B to reach the ground   [L]

18 Two particles A and B, of masses 0.4 kg and 0.3 kg respectively, are connected by a light inextensible string. The particle A is placed near the bottom of a smooth plane inclined at 30° to the horizontal. The string passes over a small smooth light pulley which is fixed at the top of the inclined plane and B hangs freely. The system is released from rest, with each portion of the string taut and in the same vertical plane as a line of greatest slope of the inclined plane. Calculate:

(a) the common acceleration, in m s^{-2}, of the two particles

(b) the tension, in N, in the string.

Given that A has not reached the pulley, find:

(c) the time taken for B to fall 6.3 m from rest

(d) the speed that B has then acquired.   [L]

19 Two particles A and B, of mass 6 kg and 1 kg respectively, are connected by a light inextensible string. Particle A is placed on a horizontal table. The string passes over a small smooth

light pulley P fixed at the edge of the table and B hangs freely. The horizontal section of the string AP is perpendicular to the edge of the table and is of length $3\,m$. The particles are released from rest with both sections of the string taut and the section PB vertical.

(i)   If the table is smooth, find:

    (a) the tension, in N, in the string

    (b) the time, in seconds to 1 decimal place, taken by A to reach the pulley.

(ii) If instead, the table is rough and the coefficient of friction between A and the table is $\frac{1}{8}$, find:

    (c) the acceleration, in $m\,s^{-2}$, of each particle

    (d) the tension, in N, in the string.   [L]

20 A car of mass $950\,kg$ tows a caravan of mass $650\,kg$ along a horizontal road. The engine of the car produces a tractive force of $2200\,N$.

(a) Find the acceleration of the car and the caravan when neither the car nor the caravan is subjected to any frictional resistance.

The car is now subjected to a constant frictional resistance of $150\,N$ and the caravan to a constant frictional resistance of $200\,N$. The engine still produces a tractive force of $2200\,N$.

(b) Show, in two *separate* diagrams, *all* the horizontal forces acting on (i) the car (ii) the caravan.

(c) Find, in $m\,s^{-2}$, the magnitude of the acceleration of the car and the caravan.

(d) Find, to the nearest N, the magnitude of the tension in the coupling between the car and the caravan.   [L]

21 Two particles A and B, of masses $0.4\,kg$ and $0.3\,kg$ respectively, are connected by a light inextensible string which passes over a smooth light fixed pulley. The particles are released from rest with the string taut and the hanging parts vertical. Calculate:

(a) the acceleration, in $m\,s^{-2}$, of A

(b) the tension, in N, in the string.

The particles continue to move in this system until the instant when they have each acquired speed $3.5\,m\,s^{-1}$. At this instant both A and B are 1.4 m above horizontal ground and the string is cut. Given that each particle now moves freely under gravity, find the difference in the times, measured from the instant when the string is cut, for A and B to reach the ground. [L]

22 (i) The unit vectors **i** and **j** are horizontal and vertically upwards respectively. A particle P is projected with velocity $(14\mathbf{i} + 21\mathbf{j})\,\mathrm{m\,s^{-1}}$ from a point O on horizontal ground and moves freely under gravity. At time T seconds after projection, P strikes the ground at the point C. Calculate:

(a) the value of T

(b) the distance OC

(c) the position vector of H, the point at which P achieves its greatest height above the ground.

Use the principle of conservation of energy to show that at an instant when P is at height 10 m above the ground the speed of P is $21\,\mathrm{m\,s^{-1}}$. [L]

(ii) Write down two assumptions which you have made about the forces on the particle during its flight.

23 A particle of mass 0.4 kg is projected vertically upwards from a point A and moves freely under gravity. Given that the kinetic energy of the particle at height 5 m above A is 2.45 J, find the speed of projection of the particle. [L]

24 The total mass of a cyclist and his machine is 140 kg. The cyclist rides along a horizontal road against a constant total resistance of magnitude 50 N. Find, in J, the total work done in increasing his speed from $6\,\mathrm{m\,s^{-1}}$ to $9\,\mathrm{m\,s^{-1}}$ whilst travelling a distance of 30 m. [L]

25 (a) A car of mass 750 kg, moving along a level road, has its speed reduced from $25\,\mathrm{m\,s^{-1}}$ to $15\,\mathrm{m\,s^{-1}}$ by the brakes which produce a constant retarding force of 2250 N. Calculate the distance covered whilst the speed is being reduced.

(b) Write down the assumptions you have made about the car in part (a) and the forces on it during its journey. [L]

26 A ball is projected from a point A on horizontal ground, with speed $14\,\mathrm{m\,s^{-1}}$ at an angle of elevation $\alpha°$, and moves freely under gravity. At the instant when the ball is at a point P,

which is 5 m above the ground, the components of velocity of the ball horizontally and vertically *upwards* are both $u\,\mathrm{m\,s^{-1}}$.

(a) By using energy considerations, or otherwise, show that $u = 7$.

(b) Obtain the value of α.

(c) Find, in seconds to 2 decimal places, the time taken by the ball to reach its greatest height above the ground.

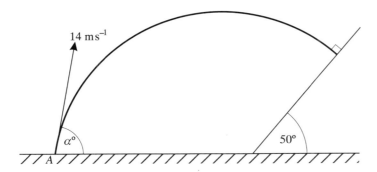

The diagram shows a fixed plane inclined at an angle of $50°$ to the horizontal. The ball strikes this plane, at right angles, with speed $V\,\mathrm{m\,s^{-1}}$.

(d) Calculate, to 1 decimal place, the value of V.   [L]

27 The unit vectors $\mathbf{i}$ and $\mathbf{j}$ are parallel to the coordinate axes Ox and Oy respectively. Two particles, P and Q, each of mass $0.4\,\mathrm{kg}$, leave the origin O simultaneously and move in the plane of Ox and Oy. The velocity vectors of P and Q at time t seconds after leaving O are $(4t\mathbf{i} + 6t\mathbf{j})\,\mathrm{m\,s^{-1}}$ and $(3t^2\mathbf{i} - 2\mathbf{j})\,\mathrm{m\,s^{-1}}$ respectively.

(a) Calculate, in J, the kinetic energy of Q at the instant when $t = 2$.

(b) Find the acceleration vector of P and hence prove that the force $\mathbf{F}$ acting on P is constant.

(c) Calculate, to the nearest $0.1\,\mathrm{N}$, the magnitude of $\mathbf{F}$ and find, to the nearest degree, the angle between Oy and the line of action of $\mathbf{F}$.

(d) Find the distance between P and Q at the instant when $t = 2$.   [L]

28 A rock-climber of mass 60 kg does 6468 J of work in slowly climbing up a fixed vertical rope.

(a) Calculate the height through which he climbs.

(b) Given that his climb takes 40 s, calculate, to the nearest watt, his average rate of working. [L]

29 A lorry is moving along a straight level road at constant speed 48 km h^{-1} against resistances of total magnitude of 900 N. Find, in kW, the rate at which the engine of the lorry is working. [L]

30 The total mass of a man and his bicycle is 80 kg. The man freewheels down a slope inclined at an angle θ to the horizontal, where $\sin \theta = \frac{1}{7}$, with constant acceleration of magnitude 0.9 m s^{-2}.

(a) Prove that the total magnitude of the resistive forces opposing the motion is 40 N.

(b) Find the time required for the man to cover 180 m from rest.

The man cycles up the same slope at constant speed 6 m s^{-1}, the resistive forces remaining unchanged.

(c) Find, in W, the power that must be exerted by the man.

(d) If the man suddenly increases his work rate by 240 W, find the magnitude of his acceleration up the slope at this instant. [L]

31 The non-gravitational resistive forces opposing the motion of a lorry of mass 2200 kg are constant and total 3100 N.

(a) The lorry is moving along a straight horizontal road at a constant speed of 20 m s^{-1}. Calculate, in kW, the rate at which the engine of the lorry is working.

(b) If this rate of working is suddenly decreased by 12 kW find, in m s^{-2}, the immediate retardation of the lorry.

(c) The lorry moves up a hill inclined at an angle θ to the horizontal, where $\sin \theta = \frac{10}{49}$, with the engine working at 90 kW. Calculate, in m s^{-1}, the greatest steady speed which the lorry can maintain up this hill. [L]

32 An engine, of mass 25 tonnes, pulls a truck of mass 10 tonnes along a railway line (1 tonne = 1000 kg). The frictional resistances to the motion of the engine and truck are constant and of magnitude 50 N per tonne mass. When the train travels horizontally the tractive force exerted by the engine is 26 kN. Calculate:

(a) the acceleration, in $m\,s^{-2}$, of the engine and truck

(b) the tension, in kN, in the coupling between the engine and the truck.

The engine and truck now start to climb a slope whose inclination to the horizontal is α, where $\sin \alpha = \frac{1}{70}$, and the frictional resistances are unaltered. At a certain instant the engine and truck are moving up the slope with speed $10\,m\,s^{-1}$ and acceleration $0.6\,m\,s^{-1}$. Calculate, at that instant:

(c) the tractive force, in kN, exerted by the engine

(d) the power, in kW, developed by the engine.   [L]

33 The engine of a car is working at a constant rate of 6 kW in driving the car along a straight horizontal road at a constant speed of $54\,km\,h^{-1}$. Find, in N, the resistance to the motion of the car.   [L]

34 The force $(6\mathbf{i} + 4\mathbf{j})$ newtons acts on a particle of mass 2 kg for 7 seconds. The particle starts from rest and at the end of the 7 seconds its velocity is $(p\mathbf{i} + q\mathbf{j})\,m\,s^{-1}$. Find the values of p and q.   [L]

35 A particle moves down a line of greatest slope of a rough plane which is inclined at 30° to the horizontal. The particle starts from rest and covers 3.5 m in time 2 s. Find the coefficient of friction between the particle and the plane.   [L]

36 A particle P, of mass 0.3 kg, is moving in a straight line on a rough horizontal floor. The speed of P decreases from $7.5\,m\,s^{-1}$ to $4\,m\,s^{-1}$ in time T seconds. Given that the coefficient of friction between P and the floor is $\frac{1}{7}$ find:

(a) the magnitude of the frictional force opposing the motion of P,

(b) the value of T.   [L]

37 (a) A concrete slab, of mass 2 kg, falls from rest at a vertical height of 10 m above firm horizontal ground from which it does not rebound. Calculate the impulse of the force exerted on the ground by the slab and state the units in which your answer is measured.   [L]

(b) State and justify briefly the assumptions you have made about the slab and the forces acting on it whilst it fell when answering part (a).

38 A coal truck, of mass 1200 kg, is moving along a straight horizontal track at $8 \, \text{m s}^{-1}$ when it collides directly with an empty truck of mass 200 kg, which is at rest but free to move, on the same track. The two trucks couple and move off with common speed $v \, \text{m s}^{-1}$. Calculate,

(a) the value of v,

(b) the loss, in J, of kinetic energy due to the collision,

(c) the impulse exerted on the stationary truck due to the collision,

(d) the constant braking force required to bring the two trucks to rest 0.25 seconds after the collision.   [L]

39 A railway truck, of mass 1500 kg and travelling with speed $6 \, \text{m s}^{-1}$ along a horizontal track, collides with a stationary truck of mass 2000 kg. After the collision the two trucks move on together, coming to rest after 12 seconds. Calculate, in N, the constant force resisting their motion after the collision. [L]

40 (a) A ball of mass 0.3 kg is released from a point at a height of 10 m above horizontal ground. After hitting the ground the ball rebounds to a height of 2.5 m. Calculate, in N s, the magnitude of the impulse of the force exerted on the ball by the ground during the impact.   [L]

(b) State two assumptions you have made about the ball and the forces acting on it in order to solve part (a).

41 A shell of mass 50 kg is fired with speed $560 \, \text{m s}^{-1}$. Given that the shell is in the barrel of the gun for $\frac{1}{20}$ second, calculate the average force, in kN, exerted on the shell by the explosive charge.   [L]

42 A pile driver of mass 350 kg drives vertically a pile of mass 630 kg into horizontal ground. The pile driver strikes the pile directly with speed $7 \, \text{m s}^{-1}$ and does not rebound, so that the pile driver and pile moves as one body into the ground.

(a) Show that the common speed of the pile driver and pile at the instant immediately after impact is $2.5 \, \text{m s}^{-1}$.

(b) Calculate, in Ns, the impulse of the force exerted by the pile driver on the pile at impact.

The pile driver and pile penetrate 0.5 m into the ground. Assuming that the force exerted by the ground on the pile and pile driver during the motion is constant,

(c) calculate, in N to 2 significant figures, the value of this constant force.   [L]

43 Two particles, A and B, of mass 0.2 kg and 0.5 kg respectively, are connected by a light inextensible string passing over a fixed smooth light pulley. The particles are released from rest with the string taut and the hanging parts vertical.

(a) Prove that the acceleration of B is of magnitude $4.2 \, \text{m s}^{-2}$.

(b) Calculate, in N, the tension in the string.

At the instant when A and B are moving with speed $6.3 \, \text{m s}^{-1}$, B strikes an inelastic horizontal floor from which it does not rebound. Calculate

(c) the magnitude of the impulse exerted on the floor by B, stating the units in which your answer is measured,

(d) the total time from the instant when the particles were first released until the instant when A first comes to instantaneous rest. (You may assume that A does not reach the pulley.)   [L]

44 Two particles P and Q, of mass 2 kg and 3 kg respectively, are placed at rest at the points A and B respectively on a smooth horizontal table. An impulse of 8 Ns, in the direction $\overrightarrow{AB}$, is given to the particle P.

(a) Calculate the speed with which particle P begins to move.

Particle P collides with particle Q. Particle P moves off with speed $v \, \text{m s}^{-1}$ and particle Q moves off with speed $2v \, \text{m s}^{-1}$, both in the direction $\overrightarrow{AB}$.

(b) Calculate the value of v.   [L]

45 A particle X, of mass 1 kg, moves with speed $3 \, \text{m s}^{-1}$ along a smooth straight horizontal groove. It strikes a particle Y, of mass 2 kg, which is at rest in the groove. Immediately after the collision Y moves with speed $2u \, \text{m s}^{-1}$ and X moves with speed $u \, \text{m s}^{-1}$ in the *opposite* direction.

(a) Calculate the value of u.

(b) Calculate, in Ns, the magnitude of the impulse received by Y at the impact. [L]

46 A particle P, of mass 0.3 kg, moves in a straight line with speed $2 \, \text{m s}^{-1}$ on a smooth horizontal plane. It collides with a stationary particle Q, of mass 0.2 kg. The two particles coalesce and move on together. Find

(a) the speed, in m s^{-1}, of the combined particle after the collision,

(b) the loss, in J, of kinetic energy due to the collision. [L]

47 A straight uniform rigid beam AB, of length 6 m and mass 15 kg, is supported at A and B and rests horizontally. A load of mass 60 kg is placed on the beam at the point C. Given that the magnitude of the force exerted on the support at A is twice the magnitude of the force exerted on the support at B, find the distance AC. [L]

48 (a) A uniform horizontal plank AB, of mass 30 kg and length 3 m, rests in equilibrium on two supports at C and D, where $AC = DB = 0.5$ m. A man of mass 76 kg stands on the plank at E, where $EB = 1$ m.
Find the magnitudes, in N, of the forces exerted by the plank on the supports at C and D. [L]

(b) How did you model the man and the plank in order to solve part (a).

49 Particles of mass 2 kg, 3 kg and p kg are placed at points whose coordinates are (1, 3), (4, 6) and (7, −8) respectively. Given that the centre of mass of these particles lies on the x-axis, find p. [L]

50 A uniform wire of length $12a$ is bent to form the sides of a triangle ABC in which $AB = 3a$, $BC = 4a$ and angle ABC is $90°$. Find the distances of the centre of mass of the triangular wire ABC from AB and BC. [L]

51 A footbridge across a stream consists of a uniform horizontal plank AB, of length 5 m and mass 140 kg, supported at the ends A and B. A man of mass 100 kg is standing at a point C on the footbridge. Given that the magnitude of the force

exerted by the support at A is twice the magnitude of the force exerted by the support at B, calculate

(a) the magnitude, in N, of the force exerted by the support at B,

(b) the distance AC. [L]

52 In $\triangle ABC$, $AB = 5a$, $BC = 12a$ and $AC = 13a$. Particles of mass m, $2m$ and $3m$ are placed at the vertices A, B and C respectively. Calculate, in terms of a, the distance of the centre of mass of the three particles from (a) side AB,

(b) side BC. [L]

53 A uniform rectangular lamina $ABCD$ is of mass $3M$; $AB = DC = 4\,$cm and $BC = AD = 6\,$cm. Particles, each of mass M, are attached to the lamina at B, C and D. Calculate the distance of the centre of mass of the loaded lamina

(a) from AB,

(b) from BC. [L]

54 Three particles of masses $0.1\,$kg, $0.2\,$kg and $0.3\,$kg are placed at the points with position vectors $(2\mathbf{i} - \mathbf{j})\,$m, $(2\mathbf{i} + 5\mathbf{j})\,$m and $(4\mathbf{i} + 2\mathbf{j})\,$m respectively. Find the position vector of the centre of mass of these particles. [L]

55 (i) A uniform lamina $ABCEF$ is obtained from a rectangle $ABCD$, with $AB = CD = 8\,$cm and $BC = AD = 6\,$cm, by removing the $\triangle EDF$, where E, F lie on CD, AD respectively, with $CE = 2\,$cm and $AF = 3\,$cm.

(a) Find the distances of the centre of mass of the lamina $ABCEF$ from AB and AD.

(b) The lamina is suspended freely from F and hangs in equilibrium under gravity. Find the angle which AF makes with the vertical. [L]

(ii) Explain what is meant by the word 'lamina'. The lamina in part (i) is described as being 'suspended freely from F'. What assumptions does this allow you to make about the forces acting on the lamina when it is suspended from F?

56

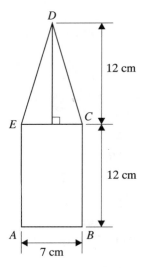

The figure shows a uniform lamina *ABCDE*; *ABCE* is a
rectangle in which *AB* = 7 cm and *BC* = 12 cm. The triangle
ECD is isosceles with *D* distant 12 cm from *EC*. Calculate the
distance of the centre of mass of the lamina from *AB*.   [L]

57   A thin uniform triangular plate *ABC* has *AB* = *AC* = 5 cm
and *BC* = 8 cm. The vertex *B* is smoothly hinged to a fixed
point and the plate hangs in equilibrium in a vertical plane.
Calculate, to the nearest degree, the angle made by *BC* with
the vertical.   [L]

58   A uniform rectangular plate *ABCD* with *AB* = 6*a* and
BC = 4*a* is of mass 6 kg. Points *H* and *E* are on the sides *BA*
and *BC* respectively such that *BH* = *BE* = 2*a*. The square
BEFH is removed from the plate.

(a) Show that the centre of mass of the remaining plate
AHFECD is at a distance:

(i) $2\frac{1}{5}a$ from *AB*

(ii) $2\frac{3}{5}a$ from *AD*.

The plate *AHFECD* is now freely suspended from the point *A*.

(b) Find, to the nearest degree, the angle between *AD* and
the downward vertical.   [L]

Examination style paper

M1

Answer all questions   **Time 90 minutes**

Whenever a numerical value of g is required, take $g = 9.8$ ms^{-2}

1 Three forces $\mathbf{F_1}$, $\mathbf{F_2}$ and $\mathbf{F_3}$ act on a particle.

$$\mathbf{F}_1 = (-6\mathbf{i} + 2\mathbf{j})\text{N}, \ \mathbf{F}_2 = (p\mathbf{i} + 4\mathbf{j})\text{N}, \ \mathbf{F}_3 = (4\mathbf{i} + q\mathbf{j})\text{N}.$$

Given that the particle is in equilibrium determine the value of p and the value of q.

(5 marks)

2 The engine of a car works at a constant rate of 16 kW. The car has a mass of 1500 kg and on a straight level road there is a constant resistance to motion of 400 N.
(a) Determine the maximum speed of the car, in m s^{-1}.
(b) Find the acceleration of the car, in m s^{-2}, when its speed is 25 m s^{-1}.

(6 marks)

3 A uniform plank AB is 6 m long and is resting in a horizontal position on two supports at A and B. The mass of the plank is 20 kg. A body of mass 30 kg is placed on the plank 1 m from A and a second body of mass 30 kg is placed on the plank 4.5 m from A. Calculate the reactions at A and B.

(7 marks)

4 Two small smooth spheres P and Q are of masses 1.8 kg and 1 kg, respectively. They move directly towards each other at speeds of 2 m s^{-1} and 6 m s^{-1}, respectively. After they collide, the direction of motion of each particle is reversed and the speed of Q is 3 m s^{-1}.

(a) Find the speed of P after the collision.
(b) Find the loss in kinetic energy due to the collision.

(9 marks)

5 At time t seconds the position vector $\mathbf{r}$ metres of a particle, P, of mass 3 kg, is given by

$$\mathbf{r} = (3t^2 + 1)\mathbf{i} + (24t - 2t^3)\mathbf{j}, \ t > 0.$$

(a) Find the speed, in m s^{-1}, of P when $t = 1$.

(b) Find the value of t when the velocity of P is perpendicular to $\mathbf{j}$.

(c) Find the magnitude and direction of the force acting on P when $t = 1$, giving the direction as an angle made with $\mathbf{i}$, to the nearest degree.

(12 marks)

6

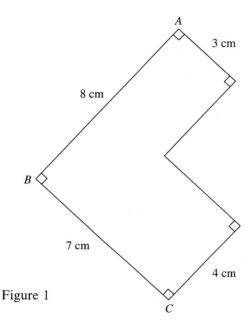

3 cm

8 cm

B

7 cm

4 cm

A

C

Figure 1

Figure 1 shows a plane figure made from a uniform sheet of metal. The shape hangs in equilibrium from the point A. The centre of mass of the shape is G.

(a) Find the distance, in cm, of G from AB.

(b) Find the distance, in cm, of G from BC.

(c) Find, to the nearest degree, the angle made by AB with the downward vertical.

(13 marks)

7

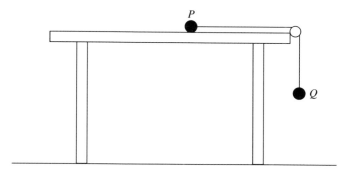

Figure 2

Figure 2 shows a particle P of mass $3\,\text{kg}$ lying on a smooth horizontal table top which is $1.5\,\text{m}$ above the floor. A light inextensible string of length $1\,\text{m}$ connects P to a particle Q, also of mass $3\,\text{kg}$, which hangs freely over a small smooth pulley at the edge of the table. Initially P is held at rest at a point $0.5\,\text{m}$ from the pulley. When the system is released from rest find

(a) the speed of P when it reaches the pulley,

(b) the tension in the string.

(c) In this problem several mathematical models have been used. Identify three of these and describe the assumptions which have been made in using these models.

(14 marks)

8

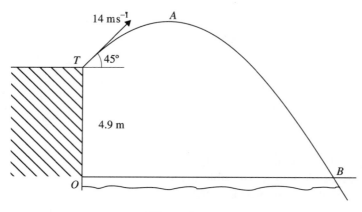

Figure 3

A golf ball is struck from the point T, at the top of a cliff 49 m above sea level, with a speed of 14 m s^{-1} at an angle of $45°$ to the horizontal, as shown in figure 3. The point O is at sea level and vertically below T. The point A is the highest point reached by the ball in its motion. The ball strikes the sea at the point B.

(a) Find the height of A above sea level.

(b) Find the distance OB.

(c) Discuss briefly the assumptions you have made in your calculation and give reasons why they are justified in this case.

(16 marks)

9 A particle P is projected up a line of greatest slope of a rough plane inclined at an angle $\arctan \frac{5}{12}$ to the horizontal. The coefficient of friction between the particle and the plane is $\frac{1}{3}$. The particle is projected from point A with a speed of 30 m s^{-1} and comes to an instantaneous rest at point B.

(a) Show that, whilst P is moving up the plane, its acceleration is of magnitude $\frac{9}{13}g$ and is directed down the plane.

(b) Find the distance AB, in metres to 3 significant figures.

(c) Find the time, in seconds to 3 significant figures, taken for P to move from A to B.

(d) Find the speed, in m s^{-1}, of the particle when it returns to A.

(18 marks)

Answers

The University of London Examinations and Assessment Council accepts no responsibility whatsoever for the accuracy or method of working in the answers given for examination questions.

Exercise 1A

1 *Car:* cost of petrol and of parking. *Assumptions:* include mileage, price of petrol, miles per gallon, availability of convenient car park.
Coach: fares, cost to and from coach station. *Assumption:* coach times convenient.
Train: fares, cost to and from station, *Assume:* convenient train available.

Possible combinations include car to bus or train station, car to free car park on outskirts plus underground train to city centre.

Refinements: could include consideration of time factor and its implications for number of meals required. Also consideration of time of day and possible delays.

2 *Relevant costs:* gas, electricity, telephone, water, Council Tax, food, transport, mortgage or rent.
Assumption: costs will be $\frac{1}{4}$ of last year's bills, and there are no unforeseen costs.
Refinements: could include more accurate estimate taking into account a particular 3 month period, and some allowance for emergencies based on past experience.

3 *Relevant facts:* length of three routes, numbers of: traffic lights, roundabouts, right turns, railway crossings.
Assumption: he drives at a constant speed, with a fixed delay for each of traffic lights, roundabouts etc.
Refinements: could include trying to obtain more accurate estimates for the delay based on trying each route for a week.

4 *Relevant facts:* distance covered in a year, cost of petrol per litre, fuel efficiency of car (km per litre), depreciation of car, servicing cost of car, M.O.T test cost, insurance cost motoring organisation (AA or similar).
Assumptions: distance covered same as last year, price of petrol is average from last year, car less efficient than last year due to age, depreciation taken as average for particular model, insurance less due to increase in no claims bonus, servicing costs as last year.
Refinements: include some allowance for unforeseen factors based on past experience. Carry out a trial to obtain better estimate of fuel efficiency in km per litre. Obtain better estimate of depreciation.

5 *Relevant factors:* costs of: hiring rooms, equipment, providing transport from nearest

station, publicity, telephone, postage, printing and secretarial help.

Assumptions: at least 100 people will attend; 50% will need transport from station; delegates will divide themselves equally between various groups.

Refinements: could obtain indication of number who will attend by using a circular. This could also give a better estimate of number requiring transport.

6 *Relevant factors:* cost of a typical shopping list, cost of transport to new supermarket.

Assumptions: prices at shop and supermarket remain basically the same; cost of transport does not vary depending on time spent on trip.

Refinements: take into account savings due to bulk purchases and own brands.

7 (a) *Relevant factors:* costs of: flying, hiring car, transport to and from airport.

(b) *Relevant factors:* costs of boat, of taking car, or cost of hiring and getting to boat.

(c) *Relevant factors:* cost of catamaran, of taking car, or cost of hiring and getting to port.

Assumptions: these are the only costs involved, and are independent of time and season.

Refinements: the travelling times for (a), (b) and (c) are different so the cost of meals needs to be taken into account. The costs of (a), (b) and (c) will vary according to time of year and time of day.

Exercise 1B

1 (a) A particle of mass 1 kg attached to one end of a light inextensible string the other end of which is attached to a fixed point. The particle moves in a vertical circle under constant gravity only.

(b) A particle of mass 0.5 kg is released from rest on a smooth plane inclined at 20° to the horizontal.

(c) A particle moves on a horizontal plane and collides with a similar particle at rest on the plane.

(d) A uniform rod of length 4 m can rotate in a vertical plane about an horizontal axis through its centre. A particle of mass 25 kg is attached at one end. A second particle of mass 20 kg is attached to the rod at the point P so that the rod rests in a horizontal position.

(e) A uniform rod rests on a rough horizontal plane with its upper end against a rough vertical wall. The rod makes an angle α with the horizontal. A particle is attached at the upper end. Find the greatest value of α for which the rod does not slip.

2 (a) This is a good model as a cricket ball is small and the air resistance is likely to be small. The variation of gravity is not great for the distances involved.

Refinements: The model may be refined by taking into account air resistance and the dependence of gravity on height.

(b) This is a good model as the puck is small and ice offers little frictional resistance to motion.

Refinements: The model may be refined by taking into account friction and also air resistance.

(c) This is a reasonable model as the size of the spheres is small compared with the other lengths involved and the mass of the rod is small compared with the other masses involved.

Refinements: The model can be refined by taking into account the shape of the spheres, the mass of the rod, the variation of gravity, and air resistance.

(d) This is a reasonable model as the size of the bucket is likely to be small compared with the length of the rope. The assumption that the string is light is valid as the weight of the bucket is likely to be much greater than that of the rope. The extension of the rope is also likely to be small so that the string may be taken as inextensible. The assumption that the pulley is smooth (its bearings are completely without friction) is likely to be reasonable if the equipment is well maintained.

Refinements: Many refinements are possible including replacement of the particle by a mass of finite size, replacement of the inextensible string by one which does extend. In addition the friction due to the bearings of the pulley and the mass of the rope could be taken into account at a later stage.

(e) This is not a very good model as the assumption that the pole may be replaced by a rigid rod neglects a fundamental property of the pole – its flexibility.

Exercise 2A

1. (a) 17 km, 13 km (b) 17 km, 13 km
   (c) 17 km, 12.2 km (d) 17 km, 16.0 km
   (e) 37 km, 28.8 km
2. (a) 7 km, 6.57 km (b) 7 km, 6.48 km
   (c) 7 km, 6.57 km (d) 7 km, 6.74 km
   (e) 9 km, 7.33 km
3. (a) 9.43 km, 032° (b) 9.43 km, 058°
   (c) 18.9 km, 032°
4. (a) 9.22 km, 091° (b) 9.22 km, 099°
   (c) 18.4 km, 099°
5. (a) 8.4 km, 155° (b) 7.7 km, 150°
   (c) 10.7 km, 167°
6. (a) $1.12 \, \text{m s}^{-1}$, 26.6° (b) $1.58 \, \text{m s}^{-1}$, 18.4°
   (c) $2.24 \, \text{m s}^{-1}$, 63.4°

Exercise 2B

1. (a) $\frac{1}{2}\mathbf{a}$ (b) $-\frac{1}{2}\mathbf{b}$ (c) $\mathbf{b} - \mathbf{a}$ (d) $\mathbf{b}$ (e) $-\mathbf{b}$
   (f) $-\frac{1}{2}\mathbf{b}$ (g) $\frac{1}{2}(\mathbf{b} - \mathbf{a})$ (h) $\mathbf{a} + \mathbf{b}$
   (i) $-(\mathbf{a} + \mathbf{b})$ (j) $\frac{1}{2}\mathbf{a} + \mathbf{b}$.
2. (a) $\mathbf{b}$ (b) $-\mathbf{a}$ (c) $\mathbf{a} + \frac{1}{2}\mathbf{b}$ (d) $-\frac{1}{2}\mathbf{a} + \mathbf{b}$
   (e) $\frac{1}{2}\mathbf{a} - \mathbf{b}$ (f) $-\frac{1}{2}\mathbf{a}$ (g) $\frac{1}{2}(\mathbf{a} - \mathbf{b})$
   (h) $\frac{1}{2}(\mathbf{b} - \mathbf{a})$ (i) $\mathbf{b} - \mathbf{a}$ (j) $\frac{1}{2}\mathbf{b} - \mathbf{a}$
3. (a) $\mathbf{a} + \mathbf{b}$ (b) $-\mathbf{a} + \mathbf{b}$ (c) $-\frac{1}{2}\mathbf{a} + \mathbf{b}$
   (d) $-\frac{1}{2}\mathbf{a} - \mathbf{b}$ (e) $\frac{1}{2}\mathbf{a} + \mathbf{b}$ (f) $\frac{1}{2}(\mathbf{a} + \mathbf{b})$
   (g) $-\mathbf{a} - \frac{1}{2}\mathbf{b}$ (h) $-\frac{1}{2}(\mathbf{a} + \mathbf{b})$ (i) $-\frac{1}{2}\mathbf{b}$
   (j) $-\mathbf{b}$
4. (a) $\overrightarrow{AO}$ (b) $\overrightarrow{OE}$ or $\overrightarrow{BO}$ (c) $\overrightarrow{OC}$ (d) $\overrightarrow{OD}$
   (e) $\overrightarrow{DO}$.
5. (a) $-\mathbf{a}$ (b) $\mathbf{b}$ (c) $\mathbf{c}$ (d) $\mathbf{a} + \mathbf{b}$
   (e) $\mathbf{a} + \mathbf{b} + \mathbf{c}$ (f) $\mathbf{b} + \mathbf{c}$ (g) $-\mathbf{a} + \mathbf{c}$
   (h) $\mathbf{a} - \mathbf{c}$ (i) $-\mathbf{a} - \mathbf{b}$ (j) $-\mathbf{b} - \mathbf{c}$

Exercise 2C

1. (a) $3\mathbf{i} + \mathbf{j}$ (b) $5\mathbf{i}$ (c) $4\mathbf{i} + 3\mathbf{j}$ (d) $-3\mathbf{i} + 4\mathbf{j}$
   (e) $5\mathbf{j}$
2. (a) $-\frac{1}{3}$ (b) $-\frac{1}{2}$
3. (a) $\frac{1}{2}$ (b) 1
4. (a) (i) 13 (ii) 8.60 (iii) 8.60 (iv) 9.43
   (v) 12.2 (b) (i) 67.4° (ii) 54.5°
   (iii) $-54.5°$ (iv) $-122°$ (v) $-55°$.
5. (a) 3.61, 56.3° (b) 3.61, $-123.7°$
   (c) 7.21, 56.3° (d) 1.80, 56.3°
   (e) 18.0, -123.7.
6. (a) 3, 0° (b) 2.24, 117° (c) 4.12, 14°
   (d) 5.10, 101° (e) 6.40, $-38.7°$
7. (a) $\frac{1}{\sqrt{2}}(\mathbf{i} + \mathbf{j})$ (b) $\frac{1}{\sqrt{2}}(\mathbf{i} - \mathbf{j})$ (c) $\frac{1}{5}(3\mathbf{i} - 4\mathbf{j})$
   (d) $\frac{1}{5}(-3\mathbf{i} + 4\mathbf{j})$ (e) $\frac{1}{5}(3\mathbf{i} - 4\mathbf{j})$
8. (a) 1.73, 1 (b) 2.60, 1.5 (c) 5.20, 3
   (d) 8.66, 5
9. (i) 5, 8.66 (ii) -5, 8.66 (iii) -8.66, 5
   (iv) -5, -8.66 (v) 5, -8.66
10. (i) (a) 3,0; 0,4 (b) 3, 4 (c) 5
    (ii) (a) 3,0; 2, 3.46 (b) 5, 3.46 (c) 6.08
    (iii) (a) 2.60, 1.5; 2, 3.46 (b) 4.60, 4.96
    (c) 6.76

(iv) (a) 4.33, 2.5; 4, 6.93   (b) 8.33, 9.43
(c) 12.6
(v) (a) 4.33, 2.5; −6.93, 4   (b) −2.60, 6.5
(c) 7

Exercise 2D

1   (a) $2\mathbf{i} + 3\mathbf{j}$ (b) $-4\mathbf{j}$ (c) $2\mathbf{i} + 7\mathbf{j}$ (d) $3\mathbf{i} + 3\mathbf{j}$
    (e) $-\mathbf{i} + 3\mathbf{j}$

2

(a)
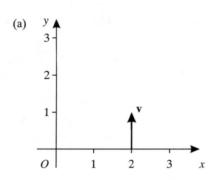

(b)

(c)

(d)
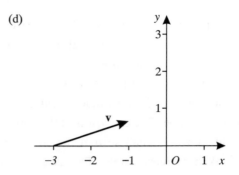

(e)
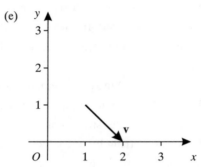

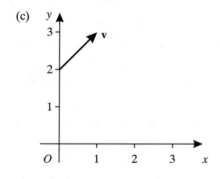

3   (a) $2\mathbf{i} + 2\mathbf{j}$ (b) $2\mathbf{i} - \mathbf{j}$ (c) $\mathbf{i} + 3\mathbf{j}$ (d) $\mathbf{j}$
    (e) $2\mathbf{i}$
4   (a) $(\mathbf{i} + 2\mathbf{j})$ (b) $(-\mathbf{i} + 2\mathbf{j})$ (c) $(\mathbf{i} + 2\mathbf{j})$
    (d) $(\mathbf{i} - \mathbf{j})$ (e) $(\mathbf{i} + 2\mathbf{j})$
5   (a) $3\mathbf{i}$; 3   (b) $3\mathbf{i} + 4\mathbf{j}$; 5   (c) $-3\mathbf{i} + 4\mathbf{j}$; 5
    (d) $(3\mathbf{i} + 4\mathbf{j}$; 5   (e) $4\mathbf{i} + 3\mathbf{j}$; $5\,\mathrm{m\,s^{-1}}$
6   (a) $5\mathbf{i}$; $5\,\mathrm{m\,s^{-1}}$ (b) $(3\mathbf{i} + 4\mathbf{j})$; $5\,\mathrm{m\,s^{-1}}$
    (c) $(5\mathbf{i} + 12\mathbf{j})$; $13\,\mathrm{m\,s^{-1}}$
    (d) $(5\mathbf{i} - 12\mathbf{j})$; $13\,\mathrm{m\,s^{-1}}$
    (e) $(8\mathbf{i} + 15\mathbf{j})$; $17\,\mathrm{m\,s^{-1}}$
7   (a) $\frac{1}{2}\mathbf{i}$ (b) $(-\mathbf{i} + 2\mathbf{j})$ (c) $(-\mathbf{i} - \mathbf{j})$
    (d) $(-4\mathbf{i} + \mathbf{j})$ (e) $(3\mathbf{i} - 5\mathbf{j})$
8   $\frac{1}{5}(3\mathbf{i} - 4\mathbf{j})$; 1
9   $20\mathbf{i} - 23\mathbf{j}$; $30\,\mathrm{m}$
10   $11\mathbf{i} + 12\mathbf{j}$; $\sqrt{265}\,\mathrm{m\,s^{-1}}$

Exercise 2E

1

(a)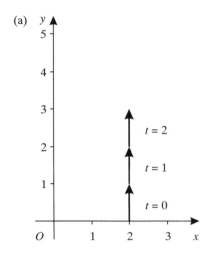

(b)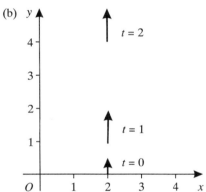

(c)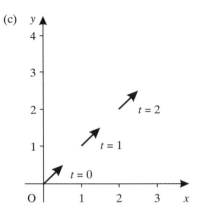

(d)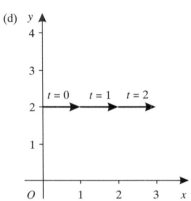

(e)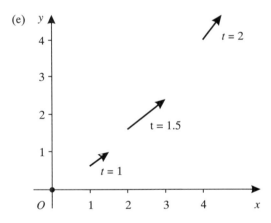

2 (a) $\mathbf{j}$; $\mathbf{j}$, 1 (b) $2t\mathbf{j}$; $2\mathbf{j}$, 2 (c) $\mathbf{i} + \mathbf{j}$; $\mathbf{i} + \mathbf{j}$, $\sqrt{2}$
 (d) $\mathbf{i}$; $\mathbf{i}$, 1 (e) $2t\mathbf{i} + \frac{3}{2}t^2\mathbf{j}$; $(2\mathbf{i} + \frac{3}{2}\mathbf{j})$, $\frac{5}{2}$

3 (a) $\mathbf{i} + 2t\mathbf{j}$; $2\mathbf{j}$ (b) $t\mathbf{i} + \mathbf{j}$; $\mathbf{i}$ (c) $3t^2\mathbf{i} + \mathbf{j}$, $6t\mathbf{i}$
 (d) $3t^2\mathbf{i}$; $6t\mathbf{i}$ (e) $t^2\mathbf{i} + t^3\mathbf{j}$; $2t\mathbf{i} + 3t^2\mathbf{j}$

4 (a) $2t\mathbf{i}$, $2\mathbf{i}$ (b) $2t\mathbf{i} + \mathbf{j}$, $2\mathbf{i}$
 (c) $2t\mathbf{i} + 2t\mathbf{j}$, $2\mathbf{i} + 2\mathbf{j}$
 (d) $3t^2\mathbf{i} + 3t^2\mathbf{j}$, $6t\mathbf{i} + 6t\mathbf{j}$

(e) $t^3\mathbf{i} + t^2\mathbf{j}$, $3t^2\mathbf{i} + 2t\mathbf{j}$

5 (a) 4 (b) 4.12 (c) 5.66 (d) 17.0 (e) 8.94

6 (a) $t^2\mathbf{i} + 2t\mathbf{j}$; $2\mathbf{i}$ (b) $t^2\mathbf{i} + t^3\mathbf{j}$; $2\mathbf{i} + 6t\mathbf{j}$
   (c) $t\mathbf{i} + t^3\mathbf{j}$; $6t\mathbf{j}$ (d) $t^4\mathbf{i} + t^3\mathbf{j}$; $12t^2\mathbf{i} + 6t\mathbf{j}$
   (e) $\frac{1}{3}t^3\mathbf{i} + 3t^4\mathbf{j}$; $2t\mathbf{i} + 36t^2\mathbf{j}$

7 (a) $(t + 1)\mathbf{i}$; $(\frac{1}{2}t^2 + t)\mathbf{i}$ (b) $\mathbf{i} + t\mathbf{j}$; $t\mathbf{i} + \frac{1}{2}t^2\mathbf{j}$
   (c) $(t + 1)\mathbf{i} + t\mathbf{j}$; $(\frac{1}{2}t^2 + t)\mathbf{i} + \frac{1}{2}t^2\mathbf{j}$
   (d) $(3t^2 + 1)\mathbf{i} + t\mathbf{j}$; $(t^3 + t)\mathbf{i} + \frac{1}{2}t^2\mathbf{j}$
   (e) $(t + 1)\mathbf{i} + 3t^2\mathbf{j}$; $(\frac{1}{2}t^2 + t)\mathbf{i} + t^3\mathbf{j}$

8

(a)

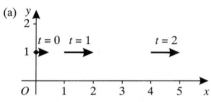

(b)

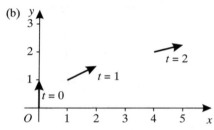

(c)

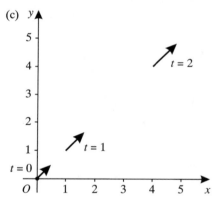

Exercise 3A

1 $15\,\mathrm{m\,s^{-1}}$ **2** $2.2\,\mathrm{m\,s^{-2}}$ **3** $17\,\mathrm{m\,s^{-1}}$
4 4.4 s **5** 18 m **6** $3.25\,\mathrm{m\,s^{-1}}$
7 1 s **8** 2 s, 28 m **9** 45 m **10** $2.5\,\mathrm{m\,s^{-2}}$

11 150 m, $1.4\,\mathrm{m\,s^{-2}}$ **12** $10\,\mathrm{m\,s^{-1}}$
13 227 m, $7.13\,\mathrm{m\,s^{-1}}$ **14** $2\,\mathrm{m\,s^{-2}}$, 21 m
15 $3\frac{1}{3}\,\mathrm{m\,s^{-2}}$, $13\frac{1}{3}\,\mathrm{m\,s^{-1}}$

Exercise 3B

1 $4.43\,\mathrm{m\,s^{-1}}$ **2** 2.02 s **3** 11.5 m
4 (a) 5.10 m (b) 2.04 s
5 (a) $25.8\,\mathrm{m\,s^{-1}}$ (b) 5.18 s
6 $10.8\,\mathrm{m\,s^{-1}}$ **7** 2.4 m **8** $15.6\,\mathrm{m\,s^{-1}}$, 3.12 m
9 2.56 s **10** 23.0 m **11** 2.47 s. Air resistance can be neglected; acceleration due to gravity is constant. Answer is slightly too small.

Exercise 3C

1 (a) $2.5\,\mathrm{m\,s^{-2}}$ (b) $1.25\,\mathrm{m\,s^{-2}}$ (c) 45 m

2

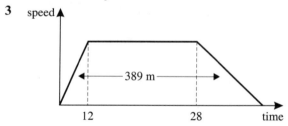

(a) 3 s (b) $6\frac{2}{3}\,\mathrm{m\,s^{-2}}$

3

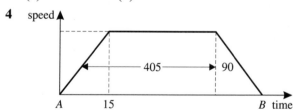

(a) $10.8\,\mathrm{m\,s^{-1}}$ (b) 56 s

4

(a) $10.8\,\mathrm{m\,s^{-1}}$ (b) 56 s

$1\,\mathrm{m\,s^{-2}}$, $1.25\,\mathrm{m\,s^{-2}}$, $46\frac{1}{2}$ s

5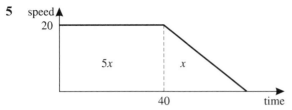

56 s

6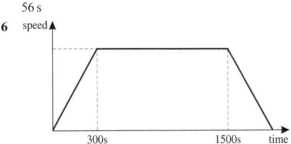

150 s, 42.8 km

Exercise 3D

1 (a) $28\,\mathrm{m\,s^{-1}}$ (b) $14\,\mathrm{m\,s^{-2}}$
2 (a) $t = 0\,\mathrm{s}, \frac{4}{9}\,\mathrm{s}$ (b) $0.132\,\mathrm{m}$
3 (a) $0\,\mathrm{m\,s^{-1}}$ (b) $-45\,\mathrm{m\,s^{-2}}$ (c) $2\,\mathrm{m\,s^{-1}}$
   (d) $8\,\mathrm{m}$
4 (a) $1\,\mathrm{m\,s^{-1}}$ (b) $\frac{2}{3}\,\mathrm{m}$ (c) $1\,\mathrm{s}$ (d) $\frac{1}{3}\,\mathrm{m}$
5 (a) $0, \frac{3}{5}\,\mathrm{s}$ (b) $4, 3.82$ (c) 17
6 (a) $5\frac{2}{3}$ (b) 2 (c) $4\frac{1}{3}$
7 $3\,\mathrm{s}$ **8** (a) $14\frac{2}{3}$ (b) $68\frac{2}{3}$.
9 $(6t - 4)\mathbf{i} + 4t\mathbf{j}$.
10 (a) $(16\mathbf{i} + 6\mathbf{j})\,\mathrm{m\,s^{-2}}$ (b) $(\frac{2}{3}\mathbf{i} + \frac{4}{5}\mathbf{j})\,\mathrm{m}$
    (c) $1.04\,\mathrm{m}$
11 (a) $(12\mathbf{i} - 8\mathbf{j})\,\mathrm{m\,s^{-1}}$ (b) $14.4\,\mathrm{m\,s^{-1}}$
    (c) $(\mathbf{i} - 2\mathbf{j})\,\mathrm{m}$ $(8\mathbf{i} - 8\mathbf{j})\,\mathrm{m}$ (d) $9.22\,\mathrm{m}$
12 $18.3\,\mathrm{m\,s^{-1}}$ **13** $3\frac{5}{6}\,\mathrm{m}$
14 (a) $((2t^3 + 3)\mathbf{i} + (t^2 + 3t + 4)\mathbf{j})\,\mathrm{m}$
    (b) $((3t + 2)\mathbf{i} + (10 - 2t)\mathbf{j})\,\mathrm{m}$ (c) $(5\mathbf{i} + 8\mathbf{j})\,\mathrm{m}$
15 (a) $((4t - a)\mathbf{i} + (3t^2 - 3)\mathbf{j})\,\mathrm{m\,s^{-1}}$ (b) $1\,\mathrm{s}, 4$
16 $((2t + 12)\mathbf{i} + 4\mathbf{j})\,\mathrm{m\,s^{-1}}, (2t\mathbf{i} + 2\mathbf{j})\,\mathrm{m\,s^{-1}}, 6\,\mathrm{s}$
17 (a) $20.6\,\mathrm{m}$ (b) $10\ \mathrm{m\,s^{-1}}, 53.1°$ below $\mathbf{i}$
    (c) $\mathbf{i}\,\mathrm{m\,s^{-2}}$

Exercise 3E

1 $3.50\,\mathrm{s}, 52.5\,\mathrm{m}$ **2** $30.6\,\mathrm{m}$ **3** $11.7\,\mathrm{m\,s^{-1}}$
4 $79.9\,\mathrm{m}$ **5** $2.67\mathrm{s}, 80.2\,\mathrm{m}$

6 $14\,\mathrm{m\,s^{-1}}, 1.23\,\mathrm{m}$
7 $(6\mathbf{i} + 13.0\mathbf{j})\,\mathrm{m}$ **8** $0, 21.6$ **9** $14\,\mathrm{m\,s^{-1}}$
10 (a) $4.43\,\mathrm{m\,s^{-1}}$ (b) $14.7\,\mathrm{m\,s^{-1}}$,
    $17.5°$ below horizontal
11 $3.68\,\mathrm{m}$ **12** $25\,\mathrm{m\,s^{-1}}, 14.7\,\mathrm{m\,s^{-1}}, 29\,\mathrm{m\,s^{-1}}$,
    $30.5°$ below horizontal **13** $(14\mathbf{i} + 4.4\mathbf{j})\,\mathrm{m}$
14 $6\,\mathrm{m\,s^{-1}}, 2\,\mathrm{s}$ **15** $123\,\mathrm{m}$ **16** $9.49\,\mathrm{m}$
17 $16.8\,\mathrm{m\,s^{-1}}, 4.97\,\mathrm{m}$ **18** $0.225\,\mathrm{m}$

Exercise 3F

1 (a) $3.54\,\mathrm{s}$ (b) $61.3\,\mathrm{m}$ (c) $7.07\,\mathrm{s}$ (d) $245\,\mathrm{m}$
2 (a) $2.86\,\mathrm{s}$ (b) $40\,\mathrm{m}$ (c) $5.71\,\mathrm{s}$ (d) $277\,\mathrm{m}$
3 $36.3°$ **4** (a) $13.4\,\mathrm{m\,s^{-1}}$,
    $38.6°$ above horizontal (b) $10.6\,\mathrm{m\,s^{-1}}$,
    $7.67°$ below horizontal
    (c) $15.4\,\mathrm{m\,s^{-1}}, 46.9°$ below horizontal.
5 $8.03\,\mathrm{m}$ **6** $7\,\mathrm{m}, 7.22\,\mathrm{m}$,
    $18.4°$ above horizontal
7 $(10\mathbf{i} - 12.4\mathbf{j})\,\mathrm{m}, 15.49\,\mathrm{m}$
8 $(8\mathbf{i} + 22.4\mathbf{j})\,\mathrm{m}, (4\mathbf{i} + 1.4\mathbf{j})\,\mathrm{m}$
9 $235\,\mathrm{m}$. Point of projection on same horizontal
    level as point of impact.
10 $7.8\,\mathrm{m\,s^{-1}}, 0.776\,\mathrm{m}$ **11** $8.64\,\mathrm{m}$
12 $35\,600\,\mathrm{m}$
13 $5.45\,\mathrm{s}, 87.1\,\mathrm{m}$ **14** $4.82\,\mathrm{m}$,
    $80.4°$ above horizontal
15 $104\,\mathrm{m}$ **16** $11.6°$ above horizontal
17 5.585 **18** $(5\mathbf{i} + 12.3\mathbf{j})\,\mathrm{m\,s^{-1}}$
19 $(10\mathbf{i} - 4.6\mathbf{j})\,\mathrm{m\,s^{-1}}, 22.5\,\mathrm{m}$
20 $4.04\,\mathrm{s}, 21.8°$ above horizontal
21 (a) $25\,\mathrm{m\,s^{-1}}$ (b) $1.6\,\mathrm{m}$ (c) $1.22\,\mathrm{m}$
22 $31.9\,\mathrm{m}, 3.12\,\mathrm{s}, 221\,\mathrm{m}$

Review exercise 1

1 (a) (i) $\mathbf{a}$ (ii) $-\mathbf{b}$ (iii) $\mathbf{b} - \mathbf{a}$ (iv) $-\mathbf{a}$
      (v) $2(\mathbf{b} - \mathbf{a})$
   (b) (i) $\overrightarrow{FE}, \overrightarrow{OD}, \overrightarrow{BC}$ (ii) $\overrightarrow{DA}$
       (iii) $\overrightarrow{BA}, \overrightarrow{CO}, \overrightarrow{OF}, \overrightarrow{DE}$
       (iv) $\overrightarrow{AB}, \overrightarrow{OC}, \overrightarrow{FO}, \overrightarrow{ED}$ (v) $\overrightarrow{CF}$.

2 (a) (i) $\mathbf{r} - \mathbf{p}$ (ii) $\mathbf{r} + \mathbf{s} - \mathbf{p} - \mathbf{q}$
  (iii) $\mathbf{p} + \mathbf{q} - \mathbf{r}$ (iv) $\mathbf{r} + \mathbf{s}$ (v) $\mathbf{r} + \mathbf{s} - \mathbf{p}$

  (b) $\mathbf{r} - \mathbf{p} = 2(\mathbf{r} + \mathbf{s} - \mathbf{p} - \mathbf{q})$

  (c) $\mathbf{r} + \mathbf{s} = k\mathbf{q}$

3

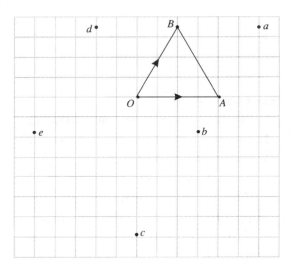

4 (a) $-2\mathbf{p}$ (b) $-\frac{1}{2}\mathbf{q}$ (c) $-\frac{3}{2}\mathbf{q}$

  (d) $\mathbf{q} - \mathbf{p}$ (e) $\frac{1}{3}(\mathbf{p} - \mathbf{q})$ (f) $-\mathbf{q}$

  (g) $\mathbf{p} - \frac{5}{2}\mathbf{q}$ (h) $-\mathbf{p} - \frac{3}{2}\mathbf{q}$ (i) $\frac{1}{3}(\mathbf{p} - 4\mathbf{q})$

  (j) $2\mathbf{p} - \mathbf{q}$

5 (a) $2\mathbf{a}$ (b) $\mathbf{a} - \mathbf{b}$ (c) $-2\mathbf{b}$

  (d) $-4\mathbf{b}$ (e) $-\mathbf{a} - 2\mathbf{b}$ (f) $-\mathbf{a} - 2\mathbf{b}$

  (g) $\frac{1}{3}(\mathbf{a} + 2\mathbf{b})$ (h) $\mathbf{a} + \mathbf{b}$ (i) $2(\mathbf{a} + \mathbf{b})$

  (j) $\mathbf{a} + 3\mathbf{b}$

6 (a) $10\,\text{s}$ (b) $(3\mathbf{i} - 4\mathbf{j})\,\text{m s}^{-1}$

7 (a) $(72\mathbf{i} + 96\mathbf{j})\,\text{km h}^{-1}$

  (b) $(144\mathbf{i} + 42\mathbf{j})\,\text{km h}^{-1}$

  (c) $(72\mathbf{i} + 96\mathbf{j})\,\text{km}, (96\mathbf{i} + 28\mathbf{j})\,\text{km}$

  (d) $72.1\,\text{km}$ (e) $(60\mathbf{i} - 95\mathbf{j})\,\text{km}$ (f) $14\ 26$

8 (a) $\frac{1}{2}$ (b) -5 (c) 24 or -24

9 (a) 12 (b) 3 (c) $\frac{1}{2}$

10 (a) $11{:}15$ (b) $245\,\text{km}$ (c) $391\,\text{km}$

11 (a) $1.12\,\text{m s}^{-2}$ (b) $31.25\,\text{s}$

12 (a) $4.2\,\text{m s}^{-1}$ (b) $5\,\text{s}$

13 (a) $45\,\text{m s}^{-1}$ (b) $2175\,\text{m}$

14 (a) $47.4\,\text{m s}^{-1}$ (b) Modelled as a particle

15 (a) $14.4\,\text{s}$ (b) $36\,\text{m s}^{-1}$

17 (a) (i) $10.5\,\text{m}$ (ii) $14.3\,\text{m s}^{-1}$

  (b) Modelled as a particle with no air
    resistance or wind and gravity is constant.

18 $2\frac{6}{7}\text{s}, 40\,\text{m}$

19 $2\,\text{s}, 4\,\text{s}$

20

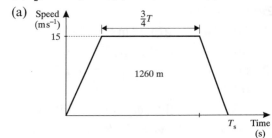

  $T = 6\text{s}$

21 (a) $\frac{1}{2}(u + 30)$ (b) $24\,\text{m s}^{-1}$ (c) $0.6\,\text{m s}^{-2}$

22 (a)

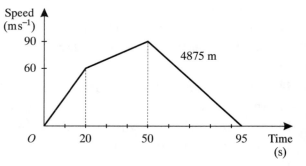

  (b) $96\,\text{s}$ (c) $\frac{15}{16}\,\text{m s}^{-2}$

23 $2475\,\text{m}, 45\,\text{s}$

24 $60\,\text{m s}^{-1}, 90\,\text{m s}^{-1}, 0\,\text{m s}^{-1}$

25 $33\,\text{m}$

26 (a) $25\,\text{m s}^{-1}$ (b) Modelled as a particle

27 $V = (2t - 4)\mathbf{i} + (3t^2 + 2at)\mathbf{j}\,\text{m s}^{-1}$
  $a = -3$

(b) $2.24 \, \text{m s}^{-1}$ 27° with positive x-axis

(c) 4 13.6 m

29 (a) $4 \, \text{m s}^{-1} (4t + 8) \text{m}$

$(6 - 2t) \, \text{m s}^{-1} (6t - t^2) \, \text{m}$

(b) 7 m   (c) $t = 5$

30 2 s

31 19 m

32 $15 \, \text{m s}^{-1}$, $19.6 \, \text{m s}^{-1}$

33 (a) 26.1 m (b) 24 m   (c) 71°

34 (a) 48 m   (b) 116 m   (c) 2.5, $20\mathbf{i} - 5.625\mathbf{j}$

35 (a) &

(b) by giving the proof

(c) $30 \, \text{m s}^{-1}$

(d) Modelled as a particle with no air resistance no wind and gravity assumed constant.

36 (a) 20.2 s   (b) 3960 m   (c) $279 \, \text{m s}^{-1}$

37 (a) (b) 40.3 (c) 62.5 (d) $39.1 \, \text{ms}^{-1}$

Exercise 4A

1 (a) 5 N, 36.9° (b) 9.43 N, 58.0°

(c) 9.43 N, 32.0° (d) 28.3 N, 32.0°

(e) 41 N, 77.3°

2 (a) 6.08 N, 25.3° (b) 6.08 N, 34.7°

(c) 8.66 N, 30° (d) 11.1 N, 68.9°

(e) 7 N, 38.2°

3 (a) 9.98 N, 56.6° (b) 9.80 N, 48.4°

(c) 14.8 N, 47.7° (d) 14.4 N, 51.2°

(e) 14.5 N, 45°

4 (a) $(10\mathbf{i} + 11\mathbf{j}) \, \text{N}$ (b) $(5\mathbf{i} + 10\mathbf{j}) \, \text{N}$

(c) $7\mathbf{i} \, \text{N}$ (d) $-6\mathbf{j} \, \text{N}$ (e) 0

5 (a) 14.9 N, 47.7° (b) 11.2 N, 63.4°

(c) 7 N, 0° (d) 6 N, 90° (e) 0

6 (a) 8.4 N, 95° (b) 3.1 N, 124°

(c) 3.5 N, 26°

Exercise 4B

1 (a) 3.46 N, 2 N (b) 3 N, 5.20 N

(c) -3 N, 5.20 N (d) -6.93 N, 4 N

(e) -5 N, -8.66 N

2 (a) 1.88 N, 0.684 N (b) 5 N, 8.66 N

(c) 12.9 N, 15.3 N

3 (a) 1 N, 1.73 N (b) 3.06 N, 2.57 N

(c) 9.66 N, 2.59 N

4 (a) 2.24 N, $-49.4°$ (b) 2.60 N, $-8°$

(c) 3.79 N, $-75.3°$

5 (a) 3.14 N, 42.5° (b) 4.40 N, 37.3°

(c) 27.5 N, 65.6°

Exercise 4C

1 (i) 5 N, 53.1° (ii) 9.43 N, 58.0°

(iii) 13 N, 67.4°

2 (i) 0.266 N, 6.36 N (ii) 0.732 N, 10.4 N

(iii) 4.93 N, 18.9 N

3 (a) (i) $x = -5$, $y = -8$ (b) $x = 1$, $y = 7$

(c) $x = -5$, $y = 6$ (d) $x = 5$, $y = 2$

(e) $x = -1$, $y = -1$

Exercise 4D

1 (a) 1.96 N (b) 0.5 kg (c) 9.8 N (d) 5 N

2 (a) 8 N (b) 21.6 N (c) 6 N (d) 1.02 kg

3 (a) 24.5 N (b) 63 N (c) 2.74 N

4 (a) 5.1 N down the line of greatest slope

(b) 0.2 N down the line of greatest slope

(c) 4.7 N up the line of greatest slope

(d) 4.8 N down the line of greatest slope

(e) 3 N down the line of greatest slope

5 (a) 17.3 N, 39 N (b) 8.66 N, 4.8 N

(c) 17.3 N, 2 N

6 (a) 0.406 N (b) 0.498 N

7 (a) 14.1 N, 14.1 N (b) 12.8 N, 12.9 N

(c) 8.82 N, 8.84 N

If α and β are not nearly equal, the peg-bag would slip on the washing line.

8 (a) 2.83 N, 2.83 N (b) 2.53 N, 2.99 N

Exercise 4E

1 (a) 2 N (b) 0.5 (c) 20 N

2 (a) 21.7 N (b) 28.9 N

3 (a) 36.5 N, 0.59 (b) 34.6 N, 0.72
4 (a) 6.70 N up the plane (b) 0.36
5 (a) 11.9 N (b) 3.00 N
6 (a) 10.5 N (b) 3.67

Exercise 5A

1 $6\,\text{m s}^{-2}$ **2** 30 N **3** 20 N **4** $5\,\text{m s}^{-2}$
5 10 N **6** 4 kg **7** $(\frac{4}{5}\mathbf{i} + \frac{6}{5}\mathbf{j})\,\text{m s}^{-2}$
8 $(8\mathbf{i} - 4\mathbf{j})\,\text{N}$ **9** $(4\mathbf{i} + \mathbf{j})\,\text{m s}^{-2}$ **10** 26 N
11 $3.75\,\text{m s}^{-2}$, 2100 N **12** $1.5\,\text{m s}^{-2}$, 3 m
13 390 N **14** 4, −2 **15** $0.64\,\text{m s}^{-2}$
16 2000 N **17** 1500 N **18** 1470 N
19 10.8 N, $(\sqrt{117}\,\text{N})$ **20** 0.0638

Exercise 5B

1 2.67 N **2** 0.372 N **3** 510 N, 510 N
4 1330 kg **5** $3.35\,\text{m s}^{-2}$ **6** 62.1 kg
7 30° **8** $6.93\,\text{m s}^{-2}$ **9** $2.49\,\text{m s}^{-2}$
10 $0.776\,\text{m s}^{-2}$ **11** $7.67\,\text{m s}^{-1}$ **12** 0.201
13 0.153 **14** 0.241 **15** 0.0827, $1.97\,\text{m s}^{-2}$
16 $1.97\,\text{m s}^{-2}$ **17** $2.35\,\text{m s}^{-2}$, 1.45 m
18 $5.20\,\text{m s}^{-2}$, $10.4\,\text{m s}^{-1}$

Exercise 5C

1 150 N, $0.25\,\text{m s}^{-2}$ **2** 786 N
3 4.44 N, $1.31\,\text{m s}^{-2}$ **4** 1820 N, 650 N
5 $966\frac{2}{3}\,\text{N}$, $483\frac{1}{3}\,\text{N}$, $733\frac{1}{3}\,\text{N}$ **6** 2500 N, 1500 N

Exercise 5D

1 $3.27\,\text{m s}^{-2}$, 26.1 N **2** $2.67\,\text{m s}^{-2}$, 99.8 N
3 $\frac{g}{3}$ **4** 3 kg, 33.6 N **5** 1.11 s, 4.67 m
6 437 N, 37.0 kg **7** $2.18\,\text{m s}^{-2}$
8 $1.09\,\text{m s}^{-2}$, 17.4 N **9** (a) $2.8\,\text{m s}^{-2}$
  (b) 4.2 N (c) $1.67\,\text{m s}^{-1}$ (d) 0.57 m
10 0.813 s **11** (a) 0.715 m (b) $2.86\,\text{m s}^{-1}$
  (c) 0.528 s **12** $1.09\,\text{m s}^{-2}$, $13.1\,\text{m s}^{-2}$,
  $0.656\,\text{m s}^{-2}$
13 0.505 kg, 3.94 N **14** (a) $0.544\,\text{m s}^{-2}$
  (b) 2.71 s (c) 2.16 m **15** 0.679, 1.05 m

Exercise 5E

1 0.84 J **2** 5.5 N **3** 588 J **4** 157 J
5 300 J **6** 12.0 m **7** 410 J **8** 274.4 J
9 0.25 **10** 24 000 J **11** 196 J **12** 567 J
13 153 J, 441 J **14** (a) 118 N (b) 4700 J
  (c) 141 000 J **15** 637 J **16** 0.0255

Exercise 5F

1 (a) 10 J (b) 4 J (c) 125 J (d) 1880 J
  (e) 108 000 J (f) 813 J
2 (a) 34.3 J gain (b) 12 700 J loss
  (c) 235 000 J gain (d) 68 600 J gain
  (e) 2160 J gain (f) 265 000 J loss
3 6 J **4** 328 000 J **5** $8\,\text{m s}^{-1}$ **6** $3\,\text{m s}^{-1}$
7 945 J **8** 80 J **9** 216 000 J
10 11.8 J, 10.0 J

Exercise 5G

1 29.4 J, $9.90\,\text{m s}^{-1}$ **2** (a) 200 J (b) 200 J
  (c) 20.4 m
3 64 J (b) 64 J (c) 4 m
4 (a) 75 J (b) 75 J (c) 3.83 m
5 (a) 7.5 J (b) 7.5 J (c) 0.425
6 $11.7\,\text{m s}^{-1}$ **7** 81.6 m **8** 20 kN
9 69.4 mm **10** (a) 37.8 J (b) 37.8 J
  (c) $5.02\,\text{m s}^{-1}$
11 1.63 m **12** 12.8 m **13** $10.8\,\text{m s}^{-1}$
14 5.31 m **15** 27 000 J, 61.2 m **16** 19.6 N
17 34.8 m **18** $16.3\,\text{m s}^{-1}$ **19** $4.23\,\text{m s}^{-1}$
20 0.138

Exercise 5H

1 12 kw **2** 9.6 kw **3** $226\frac{2}{3}\,\text{N}$ **4** $9\,\text{m s}^{-1}$
5 $12.6\,\text{m s}^{-1}$ **6** 500 N **7** (a) $1\frac{1}{3}\,\text{m s}^{-2}$
  (b) $0.296\,\text{m s}^{-2}$ (c) $20\,\text{m s}^{-1}$
8 400 N **9** 9 kw **10** $10\,\text{m s}^{-1}$
11 310 N, $0.49\,\text{m s}^{-2}$ **12** 14 kw, $6.66\,\text{m s}^{-1}$
13 $0.2\,\text{m s}^{-2}$, $11.0\,\text{m s}^{-1}$
14 $12.0\,\text{m s}^{-1}$, $0.0855\,\text{m s}^{-2}$ **15** $30.6\,\text{m s}^{-1}$

Exercise 5I

1 (a) $0.2\,\mathrm{N\,s}$ (b) $20\,\mathrm{N\,s}$ (c) $13\,500\,\mathrm{N\,s}$
  (d) $480\,\mathrm{N\,s}$ (e) $3 \times 10^6\,\mathrm{N\,s}$ (f) $13\,500\,\mathrm{N\,s}$
2 $12\,000\,\mathrm{N\,s}$ 3 $16\,200\,\mathrm{N\,s}$ 4 $2.6\,\mathrm{N\,s}$
5 $12\,\mathrm{N\,s}$ $24\,\mathrm{m\,s^{-1}}$ 6 $4.5\,\mathrm{m\,s^{-1}}$ 7 $18\,\mathrm{N}$
8 (a) $6\mathbf{i}\,\mathrm{N\,s}$ (b) $7.6\,\mathrm{m\,s^{-1}}$
9 $16\mathbf{j}\,\mathrm{N}$ 10 4 11 $(8\mathbf{i} + 10\mathbf{j})\,\mathrm{N}$
12 $5.4\,\mathrm{N\,s}$ 13 (a) $5.25\,\mathrm{N\,s}$ (b) $5.25\,\mathrm{N\,s}$
14 $30\,\mathrm{m\,s^{-1}}$ 15 (a) $7\,\mathrm{m\,s^{-1}}$ (b) $5.94\,\mathrm{m\,s^{-1}}$
  (c) $5.18\,\mathrm{N}$

Exercise 5J

1 2 2 3 3 4
4 1.6 5 3 6 4
7 3 8 1 9 4
10 3 11 $3\,\mathrm{m\,s^{-1}}$ 12 $1.5\,\mathrm{m\,s^{-1}}$, $7.5\,\mathrm{J}$
13 $6\,\mathrm{m\,s^{-1}}$, $12\,\mathrm{N\,s}$ 14 $1.04\,\mathrm{m\,s^{-1}}$
15 $4\,\mathrm{m\,s^{-1}}$, $5040\,\mathrm{J}$ 16 $500\,\mathrm{m\,s^{-1}}$
17 $200\,\mathrm{m\,s^{-1}}$, $\frac{3}{2} \times$ initial KE
18 $6.67\,\mathrm{m\,s^{-1}}$, $0.0533\,\mathrm{m}$

Exercise 6A

1 (a) $12\,\mathrm{N\,m}$ anticlockwise
6 $10\,\mathrm{N\,m}$ clockwise (c) $0\,\mathrm{N\,m}$
  (d) $15\,\mathrm{N\,m}$ clockwise (e) $10.6\,\mathrm{N\,m}$ clockwise
  (f) $3.64\,\mathrm{N\,m}$ anticlockwise (g) $0\,\mathrm{N\,m}$
  (h) $61.3\,\mathrm{N\,m}$ anticlockwise
2 (a) $14\,\mathrm{N\,m}$ anticlockwise
  (b) $6\,\mathrm{N\,m}$ clockwise (c) $2\,\mathrm{N\,m}$ anticlockwise
  (d) $1\,\mathrm{N\,m}$ anticlockwise (e) $0\,\mathrm{N\,m}$
  (f) $1.43\,\mathrm{N\,m}$ clockwise
  (g) $15\,\mathrm{Nm}$ clockwise
  (h) $5.07\,\mathrm{Nm}$ anticlockwise
3 $10\,\mathrm{N\,m}$ anticlockwise 4 $10\,\mathrm{N\,m}$ clockwise
5 $4\,\mathrm{N\,m}$ anticlockwise
6 (a) $12\,\mathrm{N\,m}$ clockwise (b) $4\,\mathrm{N\,m}$ clockwise
7 (a) $16\,\mathrm{N\,m}$ clockwise
  (b) $35\,\mathrm{N\,m}$ anticlockwise
8 (a) $4\,\mathrm{N\,m}$ anticlockwise
  (b) $21\,\mathrm{N\,m}$ clockwise

Exercise 6B

1 (a) $6\,\mathrm{N}$, $1.5\,\mathrm{m}$ (b) $5\,\mathrm{N}$, $6\,\mathrm{N}$ (c) $2\,\mathrm{N}$, $2\,\mathrm{m}$
  (d) $3\,\mathrm{N}$, $2\,\mathrm{m}$ (e) $6g\,\mathrm{N}$, $2g\,\mathrm{N}$ (f) $6g\,\mathrm{N}$, $1\,\mathrm{m}$
2 $4\,\mathrm{kg}$ 3 $2\frac{1}{9}\,\mathrm{m}$ 4 $32.7\,\mathrm{N}$, $6.53\,\mathrm{N}$
5 $30.6\,\mathrm{N}$, $28.2\,\mathrm{N}$ 6 $1.67\,\mathrm{M}$
7 (a) $100\,\mathrm{g}$, $50\,\mathrm{g}$
  (b) $\frac{2}{3}\,\mathrm{m}$. Tree trunk is unlikely to be uniform.
  If centre of mass of the tree trunk were given,
  the model would be more accurate.

Exercise 6C

1 $(4\frac{1}{6}, 0)$ 2 $(0, 4\frac{4}{7})$ 3 $(3,3)$ 4 $1\frac{2}{3}\,\mathrm{m}$
5 6 6 $7\,\mathrm{kg}$ 7 $(1, 4)$ 8 $(0.5, 0.25)$
9 (a) $1.91\,\mathrm{m}$ (b) $1.82\,\mathrm{m}$
10 (a) $4\,\mathrm{cm}$ (b) $4\frac{4}{9}\,\mathrm{cm}$
11 (a) $12\,\mathrm{cm}$ (b) $5\,\mathrm{cm}$
12 $2\,\mathrm{kg}$, $1\frac{1}{7}\,\mathrm{cm}$ 13 (a) $1.13\,\mathrm{m}$ (b) 0.467

Exercise 6D

1 (a) $(2.5, 1.5)$ (b) $(4.5, 0.5)$ (c) $(6, 6)$
  (d) $(3, 0)$ (e) $(15, 10)$ (f) $(8, 9\frac{1}{3})$
2 (a) $(2\frac{3}{7}, 2\frac{1}{14})$ (b) $(1\frac{2}{3}, 2\frac{5}{6})$ (c) $(3\frac{2}{3}, 0)$
  (d) $(7.12, 6.88)$ (e) $(1.97, 2.03)$
  (f) $(7.05, 6.05)$ (g) $(3, 2\frac{1}{5})$ (g) $(3, 3.86)$
3 (a) $0.75\,\mathrm{m}$ (b) $0.5\,\mathrm{m}$
4 (a) $10.5\,\mathrm{cm}$ (b) $2\,\mathrm{cm}$
5 (a) $1\frac{1}{6}\,\mathrm{m}$ (b) $1\,\mathrm{m}$ 6 $(2.35, -0.1)$
7 On axis of symmetry $6.26\,\mathrm{cm}$ from corner of
  square.
8 $5.42\,\mathrm{cm}$

Exercise 6E

1 $58°$ 2 $14°$ 3 $49.4°$
4 (a) $33.7°$ (b) $21.8°$ 5 $0.99\,\mathrm{cm}$ $26°$
6 (a) $6.25\,\mathrm{cm}$ (b) $5\,\mathrm{cm}$, $39°$. Peg is smooth and
  so no friction. Peg is small.
7 possible, $26.6°$
8 (a) yes
  (b) no $15.8°$

9 26.6° sliding. Surface sufficiently rough that there is no sliding for $\theta < 26.6°$

10 38.7° (a) possible (b) not possible

Review exercise 2

1 11° **2** 0.24 **3** $\frac{8}{27}$ **4** (a) 0.577
(b) the book could be modelled as a particle.

5 (a) 11.4 N (b) 13.9 N

6 (a) 80 N (b) 69.3 N **7** 0.304

8 0.44 **9** (a) −2, 7 (b) could be a particle.

10 (a) $(5 − 12\mathbf{j})\,\text{m s}^{-2}$ 13 m s^{-2} (b) 58.5 m

11 0.52 m s^{-2} **12** 3.4 m s^{-2} 144° **13** 0.58

14 1400 N **15** (a) 2.18 m s^{-2} (b) 21.8 N
(c) 4.36 m

16 (a) 3300 N (b) as particles
(c) no because all ropes are capable of being stretched when under tension.

17 (a) 4.2 m s^{-1} (b) 3.36 N (c) 2.9 m s^{-1}
(d) 0.69 s

18 (a) 1.4 m s^{-2} (b) 2.52 N (c) 35 s
(d) 4.2 m s^{-1}

19 (a) 8.4 N (b) 2.1 s (c) 0.35 m s^{-2}
(d) 9.45 N

20 (a) 3.75 m s^{-2}
(b)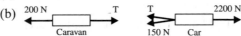
(c) 1.16 m s^{-2} (d) 952 N

21 (a) 1.4 m s^{-2} (b) 3.36 N (c) $\frac{5}{7}$ s

22 (i) (a) $4\frac{2}{7}$ s (b) 60 m (c) $30\mathbf{i} + 22.5\mathbf{j}$
(ii) Any two of: there is no air resistance, gravitational force is constant, there is no wind.

23 10.5 m s^{-1} **24** 4650 J **25** (a) $\frac{200}{3}$ m
(b) car modelled as particle, resistances negligible.

26 (b) 60° (c) 1.24 s (d) 9.14 m s^{-1}

27 (a) 29.6 J (b) $(4\mathbf{i} + 6\mathbf{j})\,\text{m s}^{-2}$
(c) 2.9 N 34° (d) 16 m

28 (a) 11 m (b) 162 W **29** 12 kW

30 (b) 20 s (c) 912 W (d) 0.5 m s^{-2}

31 (a) 62 kW (b) $\frac{3}{11}$ m s^{-2} (c) 12 m s^{-1}

32 (a) 0.693 m s^{-2} (b) 7.43 kN (c) 27.7 kN
(d) 277 kW

33 400 N

34 21, 14

35 0.37

36 (a) 0.42 N (b) 2.5 s

37 (a) 28 Ns
(b) could be modelled as a particles, gravitational force is constant, no air resistance or wind.

38 (a) 6.86 m s^{-1} (b) 5490 J (c) 1370 Ns
(d) 38400 N

39 750 N

40 (a) 6.3 Ns
(b) can be modelled as a particle, gravitational force is constant, no air resistance or wind

41 560 kN

42 (b) 1580 Ns
(c) 6100 N

43 (b) 2.8 N
(c) 3.15 Ns (d) 2.14 s

44 (a) 4 m s^{-1} (b) 1 m s^{-1}

45 (a) 1 m s^{-1}
(b) 4 Ns

46 (a) 1.2 m s^{-1} (b) 0.24 J

47 1.75 m

48 (a) 333.2 N, 705.6 N
(b) man as a particle, plank as a uniform rod

49 3

50 1.5a (a)

51 (a) 784 N (b) 0.5 m

52 (a) 6a (b) $\frac{5a}{6}$

53 (a) 3.5 cm (b) $1\frac{2}{3}$ cm **54** $(3\mathbf{i} + 2.5\mathbf{j})\,\text{m}$

55 (i) (a) $\frac{33}{13}$ cm, $\frac{58}{13}$ cm (b) 84.1°

(ii) Flat body whose thickness is negligible. Only forces are a vertical force at F and weight of lamina.

56 $9\frac{1}{3}$ cm   **57** 14°   **58** (b) 50°

Examination style paper M1

1 $p = 2, q = -6$
2 (a) $40 \, \text{m s}^{-1}$ (b) $0.16 \, \text{m s}^{-2}$
3 417 N, 368 N
4 (a) $3 \, \text{m s}^{-1}$ (b) 9J
5 (a) $19.0 \, \text{m s}^{-1}$
      (b) $t = 2$ (c) 40.2 N, $\theta = 63°$
6 (a) 2.9 cm   (b) 3.2 cm   (c) 31°

7 (a) $2.21 \, \text{m s}^{-1}$ (b) 14.7 N
      (c) particle, smooth surface, light inextensible string, smooth pulley.
8 (a) 54 m   (b) 42.9 m
      (c) assumptions – golf ball may be treated as a particle, air resistance may be neglected, g may be taken as constant – justified because ball is small, air resistance has a little effect on a small body, only slight variations in g over these heights.
9 (b) 66.3 m   (c) 4.42 s   (d) $10 \, \text{m s}^{-1}$

List of symbols and notation

The following symbols and notation are used in the London modular mathematics examinations:

{ }	the set of
$n(A)$	the number of elements in the set A
$\{x : \quad\}$	the set of all x such that
$\in$	is an element of
$\notin$	is not an element of
$\emptyset$	the empty (null) set
$\mathscr{E}$	the universal set
$\cup$	union
$\cap$	intersection
$\subset$	is a subset of
A'	the complement of the set A
PQ	operation Q followed by operation P
$f : A \rightarrow B$	f is a function under which each element of set A has an image in set **B**
$f : x \mapsto y$	f is a function under which x is mapped to y
$f(x)$	the image of x under the function f
f^{-1}	the inverse relation of the function f
fg	the function f of the function g
○——○——○	open interval on the number line
●——●——●	closed interval on the number line
$\mathbb{N}$	the set of positive integers and zero, $\{0, 1, 2, 3, \ldots\}$
$\mathbb{Z}$	the set of integers, $\{0, \pm 1, \pm 2, \pm 3, \ldots\}$
$\mathbb{Z}^+$	the set of positive integers, $\{1, 2, 3, \ldots\}$
$\mathbb{Q}$	the set of rational numbers
$\mathbb{Q}^+$	the set of positive rational numbers, $\{x : x \in \mathbb{Q}, x > 0\}$
$\mathbb{R}$	the set of real numbers
$\mathbb{R}^+$	the set of positive real numbers, $\{x : x \in \mathbb{R}, x > 0\}$
$\mathbb{R}_0^+$	the set of positive real numbers and zero, $\{x : x \in \mathbb{R}, x \geqslant 0\}$
$\mathbb{C}$	the set of complex numbers
$\sqrt{}$	the positive square root
$[a, b]$	the interval $\{x : a \leqslant x \leqslant b\}$
$(a, b]$	the interval $\{x : a < x \leqslant b\}$
(a, b)	the interval $\{x : a < x < b\}$

$\lvert x \rvert$	the modulus of $x = \begin{cases} x \text{ for } x \geqslant 0 \\ -x \text{ for } x < 0 \end{cases}, x \in \mathbb{R}$
$\approx$	is approximately equal to
$\mathbf{A}^{-1}$	the inverse of the non-singular matrix $\mathbf{A}$
$\mathbf{A}^{\mathrm{T}}$	the transpose of the matrix $\mathbf{A}$
$\det \mathbf{A}$	the determinant of the square matrix $\mathbf{A}$
$\displaystyle\sum_{r=1}^{n} \mathrm{f}(r)$	$\mathrm{f}(1) + \mathrm{f}(2) + \ldots + \mathrm{f}(n)$
$\displaystyle\prod_{r=1}^{n} \mathrm{f}(r)$	$\mathrm{f}(1)\mathrm{f}(2)\ldots\mathrm{f}(n)$
$\dbinom{n}{r}$	the binomial coefficient $\dfrac{n!}{r!(n-r)!}$ for $n \in \mathbb{Z}^{+}$ $\dfrac{n(n-1)\ldots(n-r+1)}{r!}$ for $n \in \mathbb{Q}$
$\exp x$	e^{x}
$\ln x$	the natural logarithm of x, $\log_{\mathrm{e}} x$
$\lg x$	the common logarithm of x, $\log_{10} x$
$\arcsin$	the inverse function of $\sin$ with range $[-\pi/2, \pi/2]$
$\arccos$	the inverse function of $\cos$ with range $[0, \pi]$
$\arctan$	the inverse function of $\tan$ with range $(-\pi/2, \pi/2)$
arsinh	the inverse function of $\sinh$ with range $\mathbb{R}$
arcosh	the inverse function of $\cosh$ with range $\mathbb{R}_{0}^{+}$
artanh	the inverse function of $\tanh$ with range $\mathbb{R}$
$\mathrm{f}'(x), \mathrm{f}''(x), (\mathrm{f}''')(x)$	the first, second and third derivatives of $\mathrm{f}(x)$ with respect to x
$\mathrm{f}^{(r)}(x)$	the rth derivative of $\mathrm{f}(x)$ with respect to x
$\dot{x}, \ddot{x}, \ldots$	the first, second, $\ldots$ derivatives of x with respect to t
z	a complex number, $z = x + \mathrm{i}y = r(\cos\theta + \mathrm{i}\sin\theta) = r\mathrm{e}^{\mathrm{i}\theta}$
$\operatorname{Re} z$	the real part of z, $\operatorname{Re} z = x = r\cos\theta$
$\operatorname{Im} z$	the imaginary part of z, $\operatorname{Im} z = y = r\sin\theta$
z^{*}	the conjugate of z, $z^{*} = x - \mathrm{i}y = r(\cos\theta - \mathrm{i}\sin\theta) = r\mathrm{e}^{-\mathrm{i}\theta}$
$\lvert z \rvert$	the modulus of z, $\lvert z \rvert = \sqrt{(x^2 + y^2)} = r$
$\arg z$	the principal value of the argument of z, $\arg z = \theta$, where $\left.\begin{array}{l}\sin\theta = y/r \\ \cos\theta = x/r\end{array}\right\} -\pi < \theta \leqslant \pi$
$\mathbf{a}$	the vector $\mathbf{a}$
$\overrightarrow{AB}$	the vector represented in magnitude and direction by the directed line segment AB
$\hat{\mathbf{a}}$	a unit vector in the direction of $\mathbf{a}$
$\mathbf{i}, \mathbf{j}, \mathbf{k}$	unit vectors in the directions of the cartesian coordinate axes
$\lvert \mathbf{a} \rvert$	the magnitude of $\mathbf{a}$
$\lvert \overrightarrow{AB} \rvert$	the magnitude of $\overrightarrow{AB}$
$\mathbf{a}.\mathbf{b}$	the scalar product of $\mathbf{a}$ and $\mathbf{b}$
$\mathbf{a} \times \mathbf{b}$	the vector product of $\mathbf{a}$ and $\mathbf{b}$

A'	the complement of the event A
$P(A)$	probability of the event A
$P(A\|B)$	probability of the event A conditional on the event B
$E(X)$	the mean (expectation, expected value) of the random variable X
X, Y, R, etc.	random variables
x, y, r, etc.	values of the random variables X, Y, R, etc.
$x_1, x_2 \ldots$	observations
$f_1, f_2, \ldots$	frequencies with which the observations $x_1, x_2, \ldots$ occur
$p(x)$	probability function $P(X = x)$ of the discrete random variable X
$p_1, p_2, \ldots$	probabilities of the values $x_1, x_2, \ldots$ of the discrete random variable X
$f(x), g(x), \ldots$	the value of the probability density function of a continuous random variable X
$F(x), G(x), \ldots$	the value of the (cumulative) distribution function $P(X \leqslant x)$ of a continuous random variable X
$\text{Var}(X)$	variance of the random variable X
$B(n, p)$	binomial distribution with parameters n and p
$N(\mu, \sigma^2)$	normal distribution with mean μ and variance σ^2
μ	population mean
σ^2	population variance
σ	population standard deviation
$\bar{x}$	sample mean
s^2	unbiased estimate of population variance from a sample,

$$s^2 = \frac{1}{n-1}\sum(x - \bar{x})^2$$

ϕ	probability density function of the standardised normal variable with distribution $N(0, 1)$
Φ	corresponding cumulative distribution function
α, β	regression coefficients
ρ	product-moment correlation coefficient for a population
r	product-moment correlation coefficient for a sample
$\sim p$	not p
$p \Rightarrow q$	p implies q (if p then q)
$p \Leftrightarrow q$	p implies and is implied by q (p is equivalent to q)

Index

This book is to be returned on or before the last date below.

27. JUN. 1997

0 3 NOV 1999

0 3 APR 2000

12 JUN 2002

512·5

Hebborn, John

Mechanics:No. 1